CW01023618

The Role of Social Science in the Education of Professional Practitioners

The Role of Social Science in the Education of Professional Practitioners explores the inter-relation between the social sciences and professional practice, particularly in areas of health and social welfare, and the form that professional education takes. The key issue of who provides for our health and welfare needs in the community is considered, as are the values and education that drive those people to give service in society, and how those professionals can come to a full and open understanding of their role. It focuses on the value orientation, identity development and sense of self that will enable practitioners to develop their understanding of clients' needs in the community.

The book is divided into chapters that consider the educational and learning theories that underpin professional education and how those ideas have shaped the development of the curriculum for professional practice education. Astley provides an in-depth discussion of the nature of professional roles and how the making and taking of those roles is historically influenced by politics and policy making. The nature and dynamics of the communities in which we live, including who has power, are addressed, with special reference to how the health and social welfare needs of citizens in those communities is assessed, planned for and provided.

This book will be vital reading for academics and professionals in the fields of health and social care professions' education, social and behavioural sciences, higher education, professional development and identity formation.

John Astley is a freelance consultant in education, training and organisational change.

The Role of Social Science in the Education of Professional Practitioners

The Role of Social Science in the Education of Professional Practitioners

John Astley

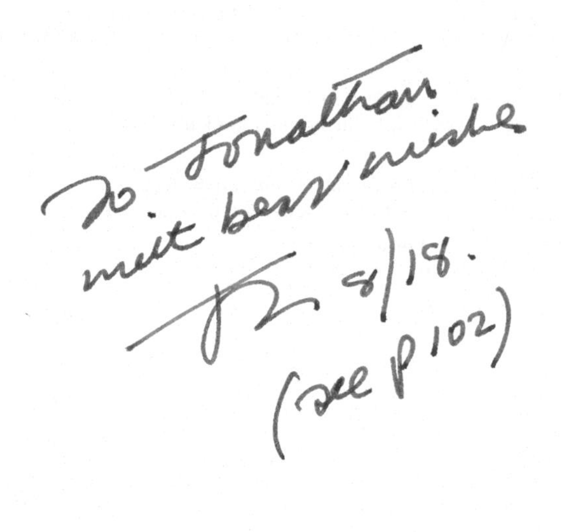

LONDON AND NEW YORK

First published 2018
by Routledge
2 Park Square, Milton Park, Abingdon, Oxon OX14 4RN

and by Routledge
711 Third Avenue, New York, NY 10017

Routledge is an imprint of the Taylor & Francis Group, an informa business

British Library Cataloguing-in-Publication Data
A catalogue record for this book is available from the British Library

Library of Congress Cataloging-in-Publication Data
A catalog record for this book has been requested

ISBN: 978-1-138-49658-3 (hbk)
ISBN: 978-1-351-02114-2 (ebk)

Typeset in Bembo
by Apex CoVantage, LLC

Printed and bound in Great Britain by
TJ International Ltd, Padstow, Cornwall

This book is dedicated to Rachel and Chris Hills, two consummate professionals, community activists and good people.

Contents

Preface

The origins of this book lie in a talk I gave at a conference on the teaching of social sciences in 2008. I have regularly returned to that first paper, largely as a consequence of continuing to think about my own role in the education of professional practitioners after 2008. I discussed with friends and colleagues the possibility of writing something much longer on these issues, and I must say they were all enthusiastic that it should be written, so long as I did it!

Acknowledgements

Several friends and colleagues have helped with this book, and I thank them all for contributing.

I would mention particularly Mel Burns, Rachel Hills, David James, Barry Lacey, Chris Mann, Jonathan Parsons and Tim Tod.

I would also thank Charlie Cherry with his help with IT, Rachel Endacott for her secretarial services, and as always Finola O'Leary for her support.

Chapter 1

An introduction

My aim in this essay is to explore what form the education of professional practitioners takes. As the title of the essay confirms my focus is on the role that social science has played and can play in these educational processes. As a sociologist I will be concentrating on the contribution of sociology and how this discipline can help us to make sense of ourselves as professional practitioners.

> In my view sociology is a symbiotic science. Its promise is to give back to people what it takes from them. This is true of all culture but sociology more than any other discipline promises to make this a practical truth.
>
> (O'Neill 1972 p. 7)

I wanted to include this quote from O'Neill's *Sociology as a Skin Trade* because it sums up well one of the key questions that should be posed to all aspiring welfare-oriented professional practitioners when they begin their educational programme: 'What is the nature of the social upon which we are working?' Social scientists are of course well situated to both pose and address this question. Not for us are the ideologically driven pronouncements of the state bureaucracy, or the blindingly obvious and narrow motives of the business world, or the 'fish and chip paper' rhetoric of the media. Our approaches to these issues are founded in reason!

However, in the post-Enlightenment world, the era of the rational and the age of reason, it has been commonplace to see the academic disciplines that make up the social sciences as purely existential, and an individual activity. Fortunately, however, we have been saved by education. In my case as a sociologist I have spent the last decades doing my best to promote an educational, and vocational, role for sociology. As a consequence there have been several of us, including I hope many of my students, who have avoided journeying down the cul-de-sac of intellectual irrelevance, hoping indeed to avoid the fate of the American sociologist Talcott Parsons, who according to his 1950s contemporary, C. Wright Mills, was always 'packing his bags but never making the journey!'

The important aspect of the journey I have made in many years as an educationalist has been to engage my many and diverse students in the pleasures of action research, as argued by O'Neill, cited above. The 'action' dimension is certainly about reflection in action, i.e. the process of research is a dialectical one in that thinking, questioning and responding goes on during the time scale of the research process. Being reflexive, reflecting and taking action are constant themes in this essay. Essentially the research process begins and ends with the focus on the wider public domain needs of those who are the 'subjects' of the research. In working with the members of a 'community' sociologists bring their general knowledge and understanding of key factors, previous research and so on, of those matters that concern the local populace. Collaboration can then determine the nature of the research process; priorities, time scales, means and ends, key agents for (and against) desired change. This should be an educational, practical, liberating and transformational process. For the professional social scientist/researcher there is virtue in this practice, and I will be discussing this key issue as an aspect of 'making sense of ourselves' later in this essay.

Not the least of the issues raised will be about practicing in the public sector, where a life's work is often more purpose than personal achievement. There are crucial links here with ideas about empathy and a utopian vision for the 'good society'. Those of us reflecting on role performance in professional practice recognise how empathy can create a space in thought and deed where seeing beyond the status quo becomes feasible.

> Professional practitioners play a range of key roles that touch all our lives. This is particularly true of those practitioners involved in welfare provisions in the widest sense. For some members of society the interventions that these practitioners make are an aspect of the significant transitions that they, as voluntary or involuntary clients, are making in their life cycle.
>
> (Astley 2006 p. 5)

There are several places in this essay where the idea of 'civil society' is addressed as an aspect of our general thinking about the form of the society in which we all live and practice. It would be useful therefore to offer a brief definition at this point of my discussion. Civil society is usually defined as a society considered as a community of citizens linked by common interests and collective activity. Clearly this is true of life in the UK, but also, depending on any one person's or cultural group's perspective, decidedly wide of the mark. This standard and paradigmatic definition sits alongside equally dominant ideas about the pluralist state that benignly presides over us all and takes care of our security and our needs. Both these ideas are a key aspect of the post-war (1945) 'settlement' between the interests of labour, capital and the state, where a plurality of power exists between these vested interests. The people, those citizens of civil society make regular choices of representatives to sit in the legislature, the law-making body that shapes and protects our collective interests. The post-war settlement, ushered in by the Labour government in 1945 was cemented together by a 'welfare state' to replace the necessary 'warfare state'. The 'deal' was that certain guarantees would be given to the people, as citizens, for example, full employment and access to a range of health and welfare provisions the like of which the majority of people in the UK had never experienced before. Professionals were widely seen as one key element to this re-balancing of access and opportunity because of their role as experts in their respective fields, and in place to give service to the whole of civil society without fear or favour.

The role of professionals, and the various claims to the equanimity in everyday life, has over the following decades been closely watched and argued about. There are several places in this essay where this pluralistic explanation of the management of society is questioned and the many changes to the roles and performance of professional practitioners examined.

Before looking deeper at the issue of curriculum development with regard to professional practice I should seek to explain my approach to contextualising the processes of thinking about designing a curriculum.

Throughout this essay I will be considering the issue of 'making sense of ourselves', which emphasises the need for professional practitioners to understand and become self-aware reflectors in action. For me the use of the word 'making' emphasises a process of development. The experimental and the experiential come into this dialectic.

I shall be exploring a range of issues concerned with curriculum, including assumptions about, and the creation of a curriculum. If we look at all (or certainly most) curricula for the education/training of professionals, we will find that social science is an integral part of that curriculum, but exactly what 'part' has changed, and is changing, will be addressed. The disciplines that tend to dominate are sociology and psychology, but economics can be found there to, for example linked in with the study of social policy. Political science, political economy and political history in one form or another will also often be linked with policy and social and public administration aspects of a curriculum. It is also the case that the generic discipline of social policy has consistently been on the professional practice curriculum, and although often in addition to sociology has been usually taught by sociologists, myself included. I should add here that there was an extraordinary growth in sociology between the mid-1960s and the mid-1970s. This happened across the whole spectrum of schooling (including plenty of social studies–type courses at CSE level) and education. The specific growth of sociology departments and lecturers in higher education reflected both an institutional enthusiasm for the subject, especially so in the new post-Robbins universities, and student demand. In addition to that, and depending on where we might look there will be (social) geographers, particularly on demographic issues, and anthropologists on cultural and ethnic diversity for example. The growth of interest within social science of a cultural time/space/globalisation/ecology dimension to the human condition is a major factor here. (See for example Anthony McGrew in Stuart Hall et al *Modernity and its Futures* 1992). This was also the era of considerable growth in national social science teaching associations and networks, and the increasing influence of research bodies like the social science research council. From the late 1970s onwards Thatcher & Co. embarked on a quest to reduce the influence of the social sciences in general and sociology in particular.

Who gets to have a piece of any curriculum development will be influenced by the membership of any specific curriculum design team in a particular place and time, which in turn will reflect both the discipline diversity of any institution and the desire of academics to actually be involved in this field of research and education. Also, any design team will almost certainly contain those people who have a specific vocational interest in the education of professional practitioners, their research and teaching bound up in a reciprocal dimension. In my experience over recent decades it is increasingly the case that the curriculum for professional education is often designed by people who have been, or still are, actually working in a specific professional practice, for example nursing or social work.

> Whose responsibility is it to communicate and implement research? Should researchers be held responsible for ensuring that the results of their work are incorporated into day-to-day professional practice? Is it an organisational responsibility for providers of health and social care . . .? Or is it perhaps the responsibility of individual practitioners to keep up to date with the latest evidence in their field and change their practice accordingly.
> (Needham 2000 p. 131)

Or all of the above? Professional 'social capital', that deep cultural reservoir of knowledge, allied to 'human capital' and a striving for possessive individualism is one possible outcome that is discussed below.

It is also the case that several of the pioneers of a more higher education–led social work training curriculum argued that these education processes should happen within a social science context. Indeed some prominent providers, for example A. H. Halsey at Oxford University, argued that social work was a social science. A supporting motif for this line of argument was the importance placed upon community work. This was particularly so in the era of 'generic' social work and creation of local authority social services departments – the Seebohm years from the late 1960s into the 1970s. The complexity of ideas and practice in this period included continuous debates about whether social work, and certainly community work, was or was not a profession.

For a very full discussion on the Oxford contribution to these developments see *Social enquiry, social reform and social action: one hundred years of Barnett House*, George Smith et al 2014. I make reference to this department's current courses later in this essay.

It is also important to suggest here that social scientists have often found great(er) virtue in designing and delivering educational programmes for professional practice, particularly so for welfare-oriented practice, than in their engagement in social policy making. Unfortunately working with, for, or hoping to influence policy makers can be very frustrating. Not the least of the issues here is the endless short-termism of politicians, in government or opposition. Politicians (and almost by definition the political parties they are part of) also exhibit an amnesiac character that for practising social scientists, with their debt to 'history' (the past in the present) and reasoned argument, can be difficult to take. I cannot be alone in the experience of attending meetings where some official and/or politician suggests that some area of policy needs inquiry and research, only to be told by people like me that there is already an extensive (and often ignored) body of top-quality research in the public domain! One aspect of this that many politicians and policy makers do not enjoy being reminded of is that we all need to consider the unintended as well as the intended consequences of policy. (For more on this vexing issue see Anthony King and Ivor Crewe, *The Blunders of Our Governments* 2013.)

It is not surprising therefore that many social scientists who are concerned enough to play a role in the social and welfare policy context of our everyday lives do so precisely because of the direct access that this practice of research-based curriculum design and delivery gives them to professional practitioners. Indeed there have been several occasions in the past that groups of social scientists designing curricula for professional practice have been accused of being subversives, 'Trojan horses' and worse, by those of a more conservative political allegiance. Education studies at the Open University in the 1970s and 80s is an example that comes to mind.

It has be to acknowledged that the further we in the UK, but with plenty of examples elsewhere in Europe, have moved away from the paradigmatic and pragmatic social democratic ideology of the 1940–60 period, that many more social scientists have decided to work in the commercial/business sector, adding to the 'soft power' dimension of those organisations in the struggle over the legitimation aspect of for-profit social science practice. Nor were/are these shifts in allegiances peculiar to social scientists; just consider for example the public service role of architects and planners in that same 1940s–60s period compared to now. In this regard I would remind the reader that many professionals, individually and collectively, have moved from a social democratic nuanced role in the public sector to one in the private sector; architects and dentists come to mind. Yet again values playing a key role, as the impact of 'the century of the self' oriented many towards the commercial nexus.

Given these changes in social science and other professional practice, there had to be a response, and later in this essay I will be examining how critical theory increasingly became an approach favoured by social scientists in general, and especially so by those of us engaged in the education of professional practitioners.

> Critical theory refuses to identify freedom with any institutional arrangement or fixed system of thought. It questions the hidden assumptions and purposes of competing theories and existing forms of practice. It has little use for what is known as "perennial philosophy". Critical theory insists that thought must respond to the new problems and the new possibilities for liberation that arise from changing historical circumstances. Interdisciplinary and uniquely experimental in character, deeply sceptical of tradition and all absolute claims, critical theory was always concerned not merely with how things were but how they might be.
>
> (Bronner 2011 p. 1/2)

And although this is a general and rather basic definition it serves to indicate where my argument is going to travel.

As a consequence of the significant changes in political ideology, there was gradually, and is now, a much more informed awareness that global capitalism, and the power of its many advocates, is part of the problem and not part of the solution for achieving a more just and fairer society. What this has meant is that approaches to education increasingly incorporate a critique of neo-liberalism and market-led educational methodologies, while giving students first-hand skills in analysis and critique, to enable self-sufficiency in their professional lives.

> Public communication has broken down to the point where we lack the means to establish an accurate account of the world as the basis for common deliberation. This breakdown is most obvious in the near-universal acceptance of the fantasy that an unrestrained private economy could be relied on as the provider of public goods. But it can be seen too in the state and the corporations' efforts to shape personality, culture and opinion, and the way private categories – character and competence – dominate the description of politics.
>
> (Hind 2010 p. 6/7)

Dan Hind's remarks are a timely reminder of the overwhelming need for 'communicative action', an open and transformational discourse, a concept developed by Jürgen Habermas, the German sociologists and 'baby' of the Frankfurt School critical theorists.

One clear context for differences of opinion over the aims and nature of professional education is related to the increasing influence of the 'social model' in the everyday practice of health and social care along with most other 'people workers' within the welfare domain. Briefly put, for now the 'social model' suggests that professional practitioners should engage with the psycho-socio-economic and cultural contexts of the everyday lives of their clients. There is also here a tendency to comprehend the 'holistic' person as someone who thinks, behaves and develops, or not, because of the nature of the society in which they live. The very form that society takes at any one time and place has a major and significant influence on people's everyday lives. Also to be considered is the acculturation process encouraging the individual social being to adopt a taken-for-granted-ness to the way things are, and reinforcing the inevitable 'snakes and ladders' nature of life and living.

A notion of the organic takes on a series of meanings here that seek to emphasise the inter-relationship between ourselves as unique individuals and the variety of cultural groupings to which we belong by choice or otherwise. Any discussion about choices, and our freedom to make them, our agency, will then inevitably require us all to embrace a concern with the nature and role of political ideologies that act as contexts to this. Discussions about ethics, empathy and empowerment follow from this perspective, and will pose serious questions for practitioners in both their value orientation to a role, and with the relationships built, or avoided, with clients in need. There are moral and legal contexts to our lives regardless of what role we play in all these processes.

This crucial issue about the conceptualisation of the practitioner-client relationship has been discussed for some time. For example in an essay on 'Social Workers: Training and professionalism' (1972). Crescy Cannan emphasises two key concerns with the focus of both practice and education for it. Firstly there is the potential danger of separate disciplines like psychology and sociology offering very divergent explanations of human actions, rather than seeking an integrated approach to conceptualisation that develops the strengths and insights of both disciplines. Besides anything else this should allow the student to assess the relative merits of different conceptual approaches, and seek in discussion an explanation and understanding that is greater in their judgement than just the sum of the parts.

A key part of my work on teaching research skills has been to make sure that students do not spend their time 'digging holes and filling them up again'. What I have attempted in my own work, and tried to pass on, is an emancipatory action research that in my judgement brings out the best in the social sciences, and takes us some way on the quest for the good community, via the proper presentation of the problem.

However, before I develop my argument it is necessary to mention here two key issues that are historically contextual. Firstly we should remind ourselves that many social scientists are focused on using their research skills to explore the nature of, and solutions for, social problems: for example, poverty, unemployment, illness, dysfunctional families or differential educational achievement. There has also been a consistently clear link between this practice goal and the development of social policy aimed at ameliorating such 'problems'. It should also be acknowledged that many/most 'social problems' are framed and presented (to the general public) ideologically. 'Problems' are talked up and down, put onto and taken off agendas, by those who have the agency, the power to do so. There are then various ways in which such research has to be approached, and once again values play a key part.

Many social scientists, but particularly sociologists, have therefore believed that this is important work for them, usually driven by their value orientation – the way in which core values direct our thinking about the purpose of our work in everyday life. They abhor such social conditions, and the prevailing discourses associated with such phenomena, and wish to address them, influence policy makers and also a wider audience. In fact there has invariably been a correlation between their humanitarian concerns, a desire to make a difference via their expertise and experience, and a belief that policy makers will take note of, or even request, the outcomes of such research.

It is also the case that many policy makers do not like the outcomes of the research that they have commissioned, and would wish to ignore it and keep it hidden.

An interesting example of this phenomenon was with the Home Office–sponsored Community Development Projects, created in the late 1960s, but actively researching 'deprived' inner city communities throughout the 1970s. The project teams produced a number of highly critical reports focusing on the accumulative contexts of 'the costs of industrial change' (1977) but also highlighted that the aim of the state was to deal with the problem of poverty by exercising more control over the poor.

> The original brief of the Community Development Programme rested on some dubious assumptions. Poverty, bad housing and so on were, it implied, residual flaws in a society that had solved all its basic problems. There was also the 'blame the victim' element in the Programme's conception: poverty and deprivation were allegedly the fault of individuals, and 'deprived areas' places where there happened to be particular concentrations of people . . . The solutions then were supposed to lie in self-help by the poor.
>
> (CDP 1977 p. 5)

The related issue of exactly what evidence flows from political action which is 'research' based and within the public eye has also been commented on by Malcolm Dean:

> [T]he potential visibility of Home Office research means there will always be close political scrutiny of (the) programme. Ultimately the research programme is managed for political ends – to enhance the reputation of the political party in government. At best it aims at the fine-tuning of policy, not challenging it, and certainly not discrediting either it or the agencies delivering it.
>
> (Dean 2012 p. 146/7)

This issue has been revisited by many different governments over many years, and little has changed because none of those policy makers has actually addressed the fundamental contexts that I cited above. This issue will be touched on later in Chapter 5.

It should also be added here that sociology in the UK in particular has always seen a lively debate about the inter-relation between science and the arts, especially so literature. In the late nineteenth and early twentieth centuries those scholars considered sociologists were often much closer in temperament to essayists and novelists, for example Dickens, George Eliot and H. G. Wells. Certainly both Richard Hoggart and Raymond Williams, both citied in this essay, were seen by many observers and analysts of 'the social and the human condition' more sociological than many 'sociologists'. With very few exceptions, the University of Sussex with its inter-disciplinary schools is one such and notable exception; most university departments were/are academic 'silos'. (For more on this issue see Halsey 2004, chapter 1.)

Belief in the power of reasoned argument is another regular theme in this essay, but I must also address the thorny issue of social science and ideology. A more traditional approach comes from the likes of Alan Ryan, an Oxford philosopher:

> An 'ideology' has come over the past century to mean a secular and political creed, and especially it has come to carry the implication that the truth of what is said to belong to an ideology is relatively unimportant, compared with its effects on those who hear it and believe it, or compared with the social origin of the creed. Unfortunately the term is not always used in this sense; sometimes it means no more than "pertaining to ideas", so that to talk of an ideology is only to talk of a set of ideas; usually, however, the talk of ideology is meant to be talk about those ideas which are selected and held for their effects on the converted and unconverted, not for their truth.
>
> (Ryan 1970)

So, not far from 'Trojan horses' again. Ryan's argument develops to suggest that with some exceptions social science cannot be seen as similar, or even equal, to the physical and natural sciences because the 'built-in' objectivity and definitely non-ideological nature of the latter render them different. It has been commonplace for this view to be taken of 'proper' science, arguing that they offer a different kind of truth, which does miss the point that the constant internal arguments of science, and the endless power struggles and paradigm shifts merely emphasise the complexities and compromises of their ideological status.

Karabel and Halsey address these issues in their consideration of ideological factors in educational systems. Interestingly they locate their argument within an argument about human capital theory, mentioned earlier, the idea that educational experience can provide the individual with the wherewithal to be a success in society. The status afforded to professional role holders within the context of a high-status professional culture has driven a desire amongst many people to job this club:

> [E]ducation can admittedly be considered a consumption good, but the benefit deriving therefrom is undoubtedly small . . . But what must further be remarked about the theory of human capital is the direct appeal to pro-capitalist ideological sentiment that resides in its insistence that the worker is a holder of capital (as embodied in his skills and knowledge) and that he has the capacity to invest (in himself).
>
> (Karabel and Halsey 1977 p. 13)

The general tenor of this argument is that schooling in particular, but for my purposes professional education in general, can be a part of the problem of the power of the values associated with possessive individualism and continued social inequalities, rather than a part of the solution. Halsey and his colleagues were making these points forty years ago, and in an era now that promotes students of any kind as consumers of a human capital oriented and commodity focus, the 'century of the self' perspective has come to pass. (And we all remain indebted to Adam Curtis for his enlightening 2002 TV series.) This is an ideological reality of our time, and therefore anyone 'taking on' the mantle of social science in particular, but any science in general, has to deal with this reality, *this* truth. The basic fact is that the truth about any phenomenon cannot be guaranteed from one source, especially so if this source seeks to exclude the human connection. Social scientists call the multi-perspective approach to evidence 'triangulation'. As a sociologist primarily concerned with the inter-relation between cultures and community, the re-rise of social history in the 1970s has been important. For example and interestingly, an enthusiasm for oral history, and local often (auto)biographical histories, added a crucial dimension to an understanding of communities. I would mention here *Talk and Social Theory* by Frederick Erickson (2004), which is a very absorbing discussion of these issues. Raphael Samuel called this whole domain 'Theatres of Memory' (1994).

I certainly believe that this renewed enthusiasm for social history and associated methodologies, reflects the science and arts tensions mentioned above. There is also the perennial arguments about the veracity of evidence, any evidence, whether so-called quantitative, big picture with lots of stats, or qualitative with its more micro account of human attitudes and behaviour. In reality, as again suggested above, most social scientists value and use both according to their research needs. I regularly access and use online sources for stats on a number of issues, and would particularly recommend the Joseph Rowntree Foundation (JRF) website to access their extensive resources. They provide very useful analysed accounts of their own research in conjunction with evidence drawn from a wide range of comparative sources.

One consequence of this for me is that values and value orientation to role holding and performance must deal with these complexities at the outset. As a sociologist I am always wary of those people who pronounce on a social phenomenon with an unwarranted certainty.

In my experience therefore, to argue for a 'purely scientific' objectivity approach from policy makers is likely to come with many disappointments as there are many flaws in this view of action taking, because for example there are the dangers of social scientists coming to believe that they have the most authoritative voice in such discourses. There is also the fundamental problem that many, if not most of the significant policy makers in society are 'politicians', who are usually driven by ideology and not reasoned argument. An example of this is the effort made by the establishment (including the monarchy) and the state apparatus to present their collective selves as ideologically neutral in relation to social policy. This is not actually the case, and remains one of the great myths of society, aided and abetted by the media. Of course there are those occasions, usually after some natural or man-made disaster, that enlightened self-interest brings about some lasting improvements in the lives of the majority of people. However, these genuine reforms are rare, and can easily be rescinded when things are tight for the rich, powerful and privileged. Most reforms over the past century or so which the state in its finest pluralistic clothing has claimed as benign intervention have in fact been due to social class struggles that had to be settled eventually.

One aspect of this prevailing situation that really angers many social scientists is the repeating of policy mistakes despite advice and warnings given, along with the regular denial of policy recommendations made by enquiry teams set up by the state to deal with previous mistakes.

As suggested earlier, one key issue here is that in any social scientific study of, and for social policy, should give equal weight to the unintended outcomes of policy as well as the intended ones. This caveat emphasises the absolute necessity for both a contextual and historical understanding of policy making and application. The sociological/social scientific response to this awareness and research orientation has often been to engage with and listen to ordinary people, seeing them as the subjects of research rather than the objects of such. This has been the equivalent of what many anthropologists describe as 'the ethnographers coming down from the colonial veranda'. Social scientists have also become more wary of, and independent from, politicians (Lindblom and Cohen 1979). They have learnt that much!

Indeed, in the 1960s and 70s there was a significant growth in the use of approaches like ethnomethodology, which put briefly is an approach that emphasises (and seeks to replicate in the research process) the meaning-making role of all social actors in their everyday life. Social scientists like Harold Garfinkel (1967) argued that 'the methods used by the people' to understand and make sense of social phenomena should be acknowledged and taken seriously by researchers. As mentioned above, there was also at this time a growth in the popularity of oral and documentary social history, and often television (for example the consistently excellent documentary work of Granada TV based in Manchester), an attempt to archive the ways in which people, in their culture groupings, expressed their take on the quotidian, making some sense of everyday life.

Despite rumours to the contrary, professional practitioners are human beings. They are also social beings; they speak, and listen, and watch, and see. They, like everyone else, have to make sense of the world and their place and roles within it. Social scientists need to assist professionals in both understanding this fundamental reality and its consequences, and sharpen their insight by becoming more skilled at the use of these observational methodologies.

Secondly, but in the same vein, i.e. how can 'we' make sense of social realities, we should remind ourselves that the first significant engagements by social scientists with social problem–related social policy making dates from the 1950s, when sociology in particular began to expand as a discipline in higher education. This was the era of increasing influence by the likes of Richard Titmuss, of the London School of Economics, and also the 'bean-counting' and utilitarian approaches to social administration. What was also prevalent was a Fabian-type view of socialism that it could be administered, even social engineered into existence. Titmuss along with Beveridge and his famous 1942 report, along with other social problem and social policy researchers, were well aware that the impact of war on the civilian population alongside a growth in the 'Dunkirk' spirit (the inevitable recourse to self-help) had signalled a widely held demand for change, for comprehensive social reform. The 1945 general election result focused the mind of policy makers and opened the door to more sociological input. (For more on this see Alan Sked and Chris Cook's *Post-War Britain: A Political History* (1993) and David Kynaston's *A World to Build* (2008).)

This socio-political phenomenon of social reforming can also be seen as an aspect of the role of positivism in the social sciences, combined with a Whig view of history and historical development, gradualism and 'social Darwinism' combined. Public and social administration – those very labels – were commonplace in the earlier days of professional practice courses in higher education. There was even a widely expressed view that these discipline titles were non-ideological and chimed with the influential view mentioned above that the road to utopia merely needed to be properly administered into reality for all. All of these characteristics and value orientations tended to form the basis of the undergraduate curriculum, which by the mid/late 1960s became a focus of campus radicalism. Many undergraduates at that time failed to see the relevance in a good deal of the social science curriculum, and wanted/demanded change. Vacation time 'free universities' became a feature of campus life, and the authoritarian top-down attitudes of most academic staff were challenged. This was especially so around the designing of the curriculum in many subject areas. There were 'special' texts that circulated, were discussed and argued over, one such example being *One Dimensional Man* by Herbert Marcuse (1964). I recall looking at a friend's copy and discovering that he had underlined almost every line!

The social Darwinism theory (often seen similar to the orthogenic theory) mentioned above offered an evolutionary prognosis for improving/civilising the working classes in order that they might take their place as citizens in a liberal democracy. Many members of the Fabian Society embraced this perspective as an aspect of the incremental but long-term transformation towards an administered socialism. Given what I have said about sociology and social policy in the immediate post-war years, it is also worth remembering that a significant number of Fabians were within the hierarchy of the Labour movement, party and governments at that time.

Within such variants of social science there was already in existence a well-established tradition of social problem–'solving' interventions into the lives of working-class neighbourhoods and family life. The origins of social work for example pre-date the post-1945 growth in academic research–focused social science. The old ways of addressing social problems – philanthropic and paternalistic in the main – and the new academic disciplines–based ways of explanation and understanding, have invariably been uncomfortable bed-fellows (Stevenson 1976).

It should also be acknowledged that the occupational basis for social class categorisation has shifted over recent years. The traditional 'blue collar' industrial working class is much smaller, while 'white collar' roles have expanded considerably. There was a major social science research investment in the late 1960s and early 1970s into these shifts in occupation and social class. The work of John Goldthorpe et al (1969) on the rise of the 'affluent worker' sought to test out the political rhetoric from Harold MacMillan and others from 1958 that 'we are all middle class now/never had it so good' and so on. This research showed that occupations may have changed, but the cultural and educational distinctions between the classes still remained. Goldthorpe along with many other social scientists sought to test out the validity of the widespread *political* and social scientific claims that in all modern industrial societies there had been convergence, a considerable narrowing in the social stratification of those societies: more open societies, greater social mobility based on educational opportunities, changes to occupational status and wealth and income disparities. The 'voice' of MacMillan was echoed in North America and in the Soviet states.

The work of Richard Sennett (1998) in the USA and the UK has emphasised these changes, and also the lack of change, while pointing to the mounting crisis of identity and opportunity amongst those workers: 'Workers are asked to behave nimbly, to be open to change on short notice, to take risks continually, to become ever less dependent on regulations and formal procedures. This emphasis on flexibility is changing the very meaning of work'.

Indeed I would go further and emphasise that the impact of this technical rationality, and inevitable lack (suppression?) of thinking about alternative explanations for everyday life, is not just associated with the workplace.

I will say more about this later, but the changing pattern of employment in the UK for example has created a growing class of service sector workers who in the main are occupying unskilled roles, and for low pay and deteriorating working conditions: more and more precarious lives with fewer options and opportunities. The dominant view of society's 'losers' was that they consistently made poor choices in their life course including their attempts to meet the norms of a liberal democracy fixated on upward social mobility. The reality is much more likely to be that the socio-economic and cultural contexts of everyday life are the key factors in a person's development, where their choices, their ability to exercise agency, are in fact very limited.

A good deal of social science has set out to challenge these ideologically driven myths about the existence of a fully 'open society' and presented for debate a body of knowledge that also seeks to challenge the myth makers themselves.

These fundamental concerns, question marks against our value orientation and our ethical capacity for empathy, also reminds us of our dual roles as academic subject specialists and teachers. This duality of roles is not without its dangers, but is also full of rewards in dealing with the tensions that arise between these roles, hopefully pulling us towards the democratic and virtuous values inherent in our practice. However, it should be acknowledged that the 'closer' we get to our students, the more likely we are as tutors to experience emotional labour. 'Open door' attitudes to students' everyday needs and well-being can, and in my experience do, invite some tensions between the objectivity of delivering a curriculum and the subjectivity of 'being there' for adults engaged in the learning process.

Not the least of issues here is an ongoing discussion around the concept of andragogy, the methods and principles used in adult education, and directly linked to theories about *how* adults learn. Both these issues are in turn an aspect of the developing ideas about 'life-long learning'. Over many years I have experienced well-meaning tutors and curriculum designers in the field of professional practice who treat their students as children, and not as the sophisticated and life-hardened adults that most are – even adults as autodidacts and self-directed learners. As adult education tutors, our values and ethical approach are regularly tested in this respect. As curriculum designers and deliverers we need to make sense of these educational specificities and processes, of the interactions that could and do happen, and be aware of the fact that our students should not be passive recipients of the bodies of knowledge we lay before them for discussion and debate. I will be returning later, in Chapter 3, to a fuller discussion about andragogy.

As an aspect of curriculum design and content, emphasising the differences in academic disciplines can also help to raise students' awareness of what they actually know. An example that comes to mind was my co-writing and delivery of a course in Oxford, 'Culture, Media and Society', with two colleagues, one an artist/designer and the other a publisher. One of my specific contributions to the curriculum was to pose to students the question, 'Why is popular music so popular?' One of the strengths of this hybrid course was that it attracted students from a wide range of undergraduate disciplines, for example politics, history, anthropology, English, design, publishing, psychology and sociology. One of the tasks I set the students was to form themselves into academic subject groups, and then through discussion identify exactly what their own discipline could bring to an analysis of this question. This seminar-based work produced at least three valuable outcomes: 1) students became more aware of what they knew, and how that knowledge could be applied to a specific problem; 2) other students gained an insight into the way other disciplines could conceptualise the question, and compared/contrasted this with their own approach; and 3) encouraged a dialectical approach to thinking through the question I had posed to them, an intellectual problem for them to bring their knowledge into action. This approach also encourages the learning and use of researching and theorising skills that are valuable for all seasons.

Of course this example may not be of great interest to students of professional practice, but I would argue that all students should be encouraged to apply their subject discipline knowledge to whatever questions are posed to them. This can help students, and their teachers, to escape from the narrow confines of their course material, and to open up and explore contexts.

As discussed earlier, students in professional practice courses have consistently been encouraged to see clients as dysfunctional, with individual pathologies (or the 'problem' family), refusing to acknowledge the role of socio-economic context in people's lives. Keeping in mind of course that understanding such contexts (the bread and butter of sociology) is that these contexts have a history to them. It has been far too easy to see life from the point of view of the 'casework' practitioner rather than the client with needs of various kinds that have not been met well beyond the realm of the practitioner or organisation role. Indeed, developing an insight into the manner in which many 'clients' are consistently punished by officialdom, or increasingly so by their private sector agents, for the audacity of being in need! Amongst other things we are here in the realm of ethics and empathy, and social scientists with their insights into reciprocity as a central dimension of everyday life (often the glue for social relationships) can contribute a good deal to these discussions. In this regard discussions on the nature of social capital, that deep cultural reservoir of experience and know-how, are never far away. I will be discussing the value of social capital as a concept in my chapter on community.

Another key issue here which has to be addressed by course designers and students alike is that the 'welfare state' is a contested concept. Social democratic ideologies promoted from the outset that the state's role in the provision of care, services and benefits was a benign and progressive one. Neo-liberal ideologies have generally seen these provisions as an (unnecessary) burden on taxpayers, and an affront to the freedom of individuals to enter the marketplace and make their own choices. There has also consistently been an argument that we should ask whether in practice the 'welfare state' is part of the solution for a socially just allocation of (scarce) resources in seeking to meet people's needs, or, in fact, part of the problem. The institutional and bureaucratic 'distancing' of providers has consistently been seen as an explanation for a lack of empathy within the welfare systems. A good many professional practitioners engaged in the public sector provision of care, services and benefits do come to see themselves as agents of the state for good or for ill, and certainly given the managerialism promoted by successive neo-liberal governments, have felt that the main task of the practitioner is to keep 'on message', tick boxes, restrain empathy, reduce face-to-face contact, and get 'clients' of all varieties to conform. The role of practitioner has often moved from liberator to domesticator within one career lifespan.

> A central assumption (of this book) has been that only *collective* struggle for and over the form of social welfare services can counteract the tendency of the New Right to capitalise on the dissatisfactions consumers feel as *individuals* about inadequate or inappropriately run services.
>
> (Deacon 1983 p. 244)

But not just the 'consumers' of course:

> Those of us writing this book . . . are often termed 'professional'. We are social/community/advice/research workers. Often these types of jobs might seem as though they were above class. But our jobs have become increasingly disciplined, especially since the cuts in public expenditure which are pushing us all into positions and attitudes that are similar to those workers for private capital.
>
> (*In & Against the State* 1980)

Both of these assertions were, even then, a confirmation of a growing belief in the need to take the ever-dominating 'for profit' market forces solutions out of the equation and also to expose the false idea of choice for most people within the 'welfare state'.

I will return to this matter later by drawing on the pioneering work of C. Wright Mills and others, who attempted to highlight the complexities in conceptualising needs and public response.

To move towards closure of this introduction let me start by re-stating the key issue at the heart of this essay.

Given all the changes to the education/training/preparation for practice of professionals in recent decades, it is now the conventional wisdom that social science will be within the curriculum in some form and at some time. The weight given to the where, when and how of the social science contribution varies, and will reflect both factors external to a specific course and the particular values, specialisms and experience of the curriculum designers. Whatever social science brings to a course, it will bring an extensive body of theoretically informed knowledge.

The status of this knowledge will stand in relation to other knowledges, and there will be power struggle associated with this.

Students will be enveloped in the 'social model' and introduced to the contexts of their clients' lives in all their complexity. If these students are fortunate, they will also have tutors who seek to introduce them to the contexts and complexities of their 'future' role as a professional practitioner. What to make of the taking and performing of this or that role? These issues are addressed in this essay, and I will offer explanation and critique. I also enclose a number of case studies for illumination of issues that arise from the designing and delivery of courses. Some of these case studies will draw on my own experience, while also offering examples drawn from colleagues over a period of time, in a way all 'expert witnesses'.

As I have already raised in this essay, studying, let alone understanding, the human condition is beset with intellectual problems.

> It is hard to define what kind of an object of enquiry human nature is, and the methods and criteria by which a particular account of it might be supported or attacked. Many statements supposedly about human nature are in fact statements about particular human societies or particular human beings . . . Human nature may imply something fixed, settled, static; a permanent human essence. My own commitment . . . is to the language of potentialities or capacities, which amongst other things draws attention to the difference between actual and possible human behaviours.
>
> (Deacon 1983 p. 187/8)

This is another good reason for any attempt to make sense of ourselves.

However, what I am also addressing is the reflection-in-action, the making sense of immediate experience, the consideration of the demands of thinking through the personal consequences of becoming a professional practitioner. All roles have a public and a private dimension to them, and I will be exploring both in this essay. There is inevitably in my view a dialectical inter-relationship taking place between these role domains. Does the now and future practitioner have the character to deal with the realities and complexities of this role that they are taking on? What is going on in their head at the same time as the hustle and bustle of the public domain pushes them here and there? Is the design of their course, and those tutoring them, up to the task of mentoring and coaching them through this?

Readers will note the cumulative nature of the following chapters as I seek to develop my argument and bring examples to illuminate the ideas introduced. The chapters will all be contextual in the manner of returning to the central question posed above. It is inevitable that as a sociologist I will explore the contexts to social actions. I will be exploring the inter-relationships between the social and institutional structures that frame our everyday lives, and the influence of cultural group membership and personal biography. We are all unique human beings, but we are also social and cultural beings, and there are significant consequences that flow from that, for us as individuals and for the people we encounter within our own culture group and beyond.

> What I see around me all the time, and read about at length, is that many people, professional practitioners and otherwise, are practicing/practising their way out of the current malaise, resisting the pious humbug and political myopia, and working towards the "good community". What does concern me greatly is the continued sense of isolation that many people feel. Contemporary life has a way of fragmenting ourselves, of throwing into doubt the values and beliefs we have nourished. What needs to be done with some urgency is to reinvest people with a sense of collective commitments.
>
> (Astley 2006 p. 150)

I will be returning to 'community' issues later, but for now it is fair to say that the issues that have been addressed so far are essentially of an existential nature. The experience of professional education is going to have a bearing on the identity of the person *as* practitioner. My aim in this essay is to elaborate on these concerns.

I shall be introducing a number of case studies throughout this book; these will be of different lengths and will focus on a range of issues pertinent to a particular chapter and discussion. One thing I am trying to do with these case studies is to relate complex empirical 'data' to a theoretical perspective in order to show how 'university' courses have developed in an ongoing relation between the current social science paradigms, professional practice, and the increasingly complex range of providers.

The reader will note that the contexts of professional practice education and professional practice are consistently discussed in this essay. Typically sociologists concern themselves with both the macro and the micro contexts within which, even because of how, social action happens. These social actions, both collectively and individually, may often seem random, or common sense, but sociologists would point to the likelihood of this or that phenomenon given certain contexts. Adult learning theories – the why, when and how of adult learning – have increasingly been focused on the *contexts* of this learning. Adult learning is not context-free. Social phenomena like poverty and other public health issues experienced in children are likely to lead to underachievement in schooling and reduced opportunities in the workplace, lower pay, lower standard of living and quality of life and so on, and also for many, a lowering of expectations and lack of self-confidence.

A very common theme in sociological analysis is the inter-relation between social structure(s): the typical institutionally based factors of life, like family, schooling, locality, work and so on, and agency. Indeed agency contrasted with social structure, as agency indicates the amount of choice, decision-making capacity, in any one person's life. We may have agency as individuals, but the chances are that the socialisation process we experience will be characterised by other people's agency in our life course – for example parents, teachers, bosses, religious leaders. All socialisation throughout life comes with social controls of various kinds, and those social controllers are likely to be the people with agency cited above. Either acting 'on their own' as role holders, or as part of a cultural set of imperatives, expectations, etcetera, the desire to intervene into the lives of others is a significant factor. The key factors that affect our everyday lives are both invariable and variable, fixed and mutable by turns. There will be effects. This is an aspect of the contexts that must be kept in focus, not the least of reasons being that given the relative power, and therefore agency, that professionals have to intervene into the lives of their clients is significant. The degree of insight into this situation that a prospective professional practitioner has is likely to be linked both to the values they hold and the legitimacy they envisage in their role performance.

I have also argued that a useful way of looking at the various dimensions of everyday lives in society that have a bearing on our understanding of how we come to be the person we are, and scope for change, brings together social structures, with our membership in cultural groups, and a person's specific biography. Despite all the commonalities we share with others of the same gender, social class, ethnicity, locality and so on, no one else has lived, is living, our life. These are complex matters to be considered and understood.

Finally, here, just a brief nod in the direction of a definition, an ideal type, of the professions, which are of course a major issue in this essay.

A great deal has been said for over a century about the professions, professionalisation, de-professionalisation and professional practice. How and why the professions came to grow, acquire the status they have, accumulate so much power and even legitimacy for their role in our lives has not surprisingly been an arena of contested ideas. These issues about the ebb and flow of professional roles and performance are therefore at the heart of my discussion. This is especially so when set against the standard definition of a profession:

1 A systematic knowledge base coming from extended education and practice experience.
2 A (giving) service ethos, a value base of neutrality and objectivity.
3 Autonomy in role performance, and over the maintenance of standards and good governance.

It is still the case that the professions have a certain cachet in our kind of society. As occupational groups go, the professions, of all kinds, do have a high status, and usually consequent rewards. No surprise then that so many occupational groups still strive to be acknowledged as of professional status, and that many individuals wish to see themselves as a role holder in such a relatively high-status occupation, or even vocation. As a grammar school boy in the late 1950s, I was very aware of the occupational steer.

The reader should also note that I use the term 'health and social welfare' throughout this essay. I intend this to be a broad description of all health provisions and services, including of course primary care in all respects. Social welfare covers an even broader catchment of care, benefits and services, from social care for the elderly, to housing provision, to detached youth work. Inevitably the benefits spectrum is in itself an extensive aspect of the everyday lives of people and occupies a good deal of professional practice in one guise or another.

I have also made references to the private sector in my discussions about provisions and providers. My main focus in on what the state apparatus does, nationally and locally, and a good deal about what it does not now do. I also look at the voluntary and charity sector (often referred to by outsiders as 'the third sector'). Also I spend time considering social enterprises in all their myriad forms. However, the increased role of the private sector, from major corporations to individuals in private practice, is considerable. Some graduating professionals do choose to seek a role in the private sector, which may reflect that person's values and value orientation, and/or the best opportunity for them to get a job that meets *their* needs and aspirations. Virtually regardless of the health and social welfare sector of employment that people enter they will in the main have experienced a generic professional education course appropriate to their specialism.

In the following chapters, focused as they are more specifically on curriculum ideas and issues, on role taking and then on the nature of the communities in which professional practice happens, these core concerns about the form and nature of a profession and a professional are never far away. In my 2006 book, *Professionalism and Practice: culture, values and service*, I addressed the many strengths and weaknesses of professionals, and considered the increasing awareness within the professions, and society at large, that not all was well, that changes needed to be made. It is certainly true that some of the most valuable insights into the less than perfect role of professionals in our everyday lives came from within the health and social welfare and education sectors. This unfolding story is another key focus in this study. As a sociologist I have maintained my interest in just how cultural groups like the professions come about, for example what are the confluences of time, place, circumstance and so on that led to the creation of a particular occupational group, and the ideas that those people have about themselves *as a group*. How are they seen by 'outsiders'? Are they admired, considered functionally necessary, reviled? How do changes come about with a culture group? Are there pressures from within, or primarily from beyond the sanctuary of the group itself? When change comes, positively or negatively, how is it managed? Do some professional cultures manage these processes of change better than others? What role does education, internal and external, play in the development of ideas that are associated with change(s)? For example, can we note that within a particular professional culture the role of continual professional development actually encourages practitioners to consider changes to the focus of practice, and ask how this impinges on role taking and role performance. Most professions devote a good deal of time and effort to the maintenance of the dominant culture which characterises that group, for example through the core issue of standards, held dear and protected. However, all culture groups are under long-term pressures from the shifts taking place in general in society. Consider for example the number of occupational culture groups that *have* disappeared, or are close to consigned to the heritage museum. The costs of industrial change in the last fifty years have reduced many occupational culture groups to the status of memories, even material culture ones like buildings, or mining winding gear, or agricultural tools.

I regularly return to the importance for me of the inter-relation between the ever-shifting, but amazingly fixed, social structures of society, the symbolically loaded cultures to which people hold, to the individual biographies of those individuals who are all the essential ingredients.

There are some key themes in this essay, for example role taking and performance, reflective practice and changes to welfare needs provision. These themes are revisited in most of the chapters that follow, and should be seen by the reader as both familiar, but set in the specific context of a chapter. These are regular reminders of the direction of travel for the arguments at the heart of my essay. This is one aspect of my theorising-in-action as the arguments in this essay develop.

As a social scientist, and specifically so a sociologist, I shall be making use of some familiar concepts throughout this essay – concepts like role, values, power, community, culture, socialisation, social control, class and gender. These concepts, along with so many others, are part of my 'intellectual toolbox'; they aid me as starting points in thinking about an intellectual problem I have, how can I understand this or that social phenomenon, this event, that encounter. I have to start my thinking from somewhere and these concepts which I share with the social scientific community help to organise my thinking; they are a short-cut, signposts. However, all of these handy concepts are contested, and once we get beyond a general sense of historical agreement on these 'formula' ideas, there are inevitably many arguments over detail.

Having promoted the validity of such conceptual approaches but highlighted 'contestation', it is important to add a further word of caution in specific relation to a recurrent theme in this essay, the political dimension of policy making, and take one of my key concepts, 'values' as an example:

> Not only does the term 'values' obscure a range of meanings, but behind each meaning lies a set of differing assumptions about the nature of the policy process. 'Values in policy-making' is taken here to refer to the policy-makers' subjective understanding of the environment in which they operate. As this statement implies, policy-making is defined largely in terms of the management of forces operating within the perceived environment of the political system.
>
> (Pallitt 1979 p. 31)

And, to conclude with another 'unfolding story' which underlines, again, why this book is called *The Role of Social Science in the Education of Professional Practitioners*:

***

Journeys without end: now and then

Most of us can remember where we were at the time of a major event, the Kennedy assassination, the attack on the 'twin towers', or the tumbling of the Berlin Wall. I remember where I was when I heard of the notorious 'rivers of blood' speech by Enoch Powell in 1968, a speech through which, in the area where I was brought up, the industrial West Midlands, he achieved hero status overnight. Powell's influence quite possibly gave the Tories their unexpected election victory in 1970 when most of the marginal seats in the West Midlands swung the Tories way. I myself, and most of my working-class contemporaries who were feeling abandoned and forgotten by the political elite, were amongst Powell's cheerleaders and voted Tory in that election.

Within the next five years my political views had changed and I was diametrically opposed to the views as expressed by Powell and to the philosophy of the right in general. How did that happen? There was no epiphany moment; the journey I had made, and you might say needed to make was no overnight passage to nirvana. Change, as they say is not a destination but a journey, and my own journey had several stages. First, there was a change of jobs that thrust me into a working environment of well-educated, yes liberal-minded and dare I say it, middle-class people who offered me an alternative way of looking at the issues like immigration – to name just one.

The journey continued when I was selected for professional social work training at a well-known polytechnic, as I still like to call them. As a school leaver from secondary modern school at the age of fifteen, I was, in my late twenties given a second chance to digest learning, and I was bringing an insatiable appetite to the table. Subjects like sociology, psychology and social history were served up in generous portions by enthusiastic teachers who knew how to spice up their subject. We had law, psychology and social work methods, but the subject that had the most profound impact was social policy. The human interactive subjects that dealt with social work practice were all very interesting, but it was the social policy lectures that showed me what social workers should be doing and why.

Connections

It wasn't just the sheer range of subject matter, public health, health and social insurance, housing, education (and don't get me going on that one) to name but a few that I was learning about. It was the ideology that drove the policies forward from ideas to the statute book often, indeed almost always, after a prolonged period of struggle. If the policy was the words, then the history was the music, and I was finding out that it was only by looking at the history that we get anywhere near the full story. I was learning that old maxim that there is no growth without conflict and in the case of social reform there was conflict aplenty as the whole panoply of the state was brought to bear on those campaigning for change.

Take the series of epic battles Lloyd George had to fight with the House of Lords, and the Conservative opposition to his 1906 budget, which contained significant welfare reforms, through the House of Lords and on the statute book. We learnt that these reforms paved the way for the Beveridge report that came some forty years later and built the foundations of the modern welfare state that for a time was subject to post-war consensus. I did say for a time. Yet for all that it was the history of education policy, a history that can be described as a slow incremental process that really made me sit up and take notice. I still remember being stunned when we learnt that those same people who said it was time to educate our masters also made the case to teach the working classes to read but not write – after all, we don't want any seditious literature being spread around, do we?

There was more, lots more. Soon we were finding out about the 1834 Poor Law Amendment Act, the one that gave us the workhouses that laid within the living memory of people I knew when I was a boy. Is any social history of Britain complete without looking at the Chartist movement? I was on a roll, and in a role, and with a lot of extra input from our social policy lecturer I even found time to read up and acquire a basic understanding of economics to help me understand the financial crisis of 1976 that resulted in the Labour government borrowing $2.8 billion dollars from the International Monetary Fund.

As I began to assimilate all (perhaps I should say some) of this new information something else happened, something I wish could happen to all of those people who feel disenfranchised and abandoned by the political class. It was the connection, real, tangible, verifiable connection between parliamentary legislation and us, the people, the general population. If ever there was an epiphany moment it was this. When I think about those formative years in the West Midlands I can still hear the echo of the familiar lament:

> It doesn't matter who gets in, it won't make any difference to me.

This from those same people who send their children to school, go the doctor, receive sick pay, possibly live in a council house with sanitation and on-tap running water, have their domestic waste collected, and may even use a library. Government legislation does not make any difference to them.

Education Education Education

Whenever I reflect on my own journey I know it could not and would not have been made without the higher, formal education I was able to receive when I was given a second chance. After two years training I felt very well equipped to do social work, not just because of the intensive training in social work practice but also the social policy that forms the legal framework within which social workers operate. Everything social workers do is conferred by legal statute and when we finished our training we knew both what we were doing and why. Sadly, there have been times since then when I have not been convinced that all social workers have been sufficiently aware of this. Is there a fault in some of the social work education?

So when I think about the state of the world and how it could be made better I often think education is more important than the equitable distribution of wealth. What kind of education? Maybe I still labour under the illusion that education should be as the ancient Greeks envisioned it, where the mind was trained to develop critical, reflective faculties and above all to reason. Instead we seem to be fixated on preparing people for work, to compete in an economic bunker with little time to look up. If this is the case, then whose interests are being served by the pursuance of such a policy? Discuss.

Barry Lacey Oxford 3/18

***

Chapter 2

Thinking about curriculum theory

Before looking at the curriculum issues specific to professional education I will consider the nature and role of curriculum theory in a more general way. What I want to know about any domain of theory, in general or particular, is 'will this help?' in both my understanding of key principles, and the previous discussions about the intellectual and practical problem under consideration – in this specific instance the designing of a curriculum for the education of professional practitioners, and for its eventual effective delivery. Will this or that context of the historical ideas about the nature and role of curricula adequately guide my thinking about the curriculum design issues I will undoubtedly face either on my own, or as the member of a curriculum design team. Like most people working in higher education I have done both many times, and have inevitably developed my own style in the design process. I have often been influenced, and even constrained, by the 'house style' of the institution I was working for, or by regulations laid down by some professional standards exterior body.

Not surprisingly, the majority of curriculum theory, and continuing contemporary speculation about the appropriate approaches to curriculum styles, has focused on schooling and on pedagogy. This is an historical and world-wide phenomenon, and even if we narrow the focus to the UK in the last hundred years or so, there has been a good deal of discussion and argument. My approach in this book is to focus on andragogy, educating adults, rather than children, but many of the questions about curriculum theory are similar. So, before I look at the specific issue of professional practice education, it is valuable to consider some generalities about curriculum theory: what is it, is it important, does it matter? But also do these theories stand up to scrutiny?

> From the outset, then, there were serious discrepancies between the rationale for a philosophical approach to educational theory and the particular way in which it was put into practice. In practice, educational theory was never concerned with developing reflective and philosophical thinking in teachers, but only with presenting the summarised results of the philosophical thinking of others. Educational theorising emerged, not as a distinctive

> way of thinking in which teachers actively engaged, but as the passive digestion . . . of philosophical doctrines.
>
> (Carr and Kemmis 1986 p. 53/54)

This is a familiar complaint, and a theme I will return to several times in this essay. The general dissatisfaction with the 'gap' between abstract thought and practical application has gone further. Where is the evidence, and what are the sources of that evidence? Do they match up to even a reasonable benchmark of science and particularly for me, social science?

> This process of close critical appraisal isn't something we tolerate reluctantly, in science, with a grudge: far from it. Criticism and close examination of evidence is actively welcomed – it is the absolute core of the process – because ideas only exist to be pulled apart, and this is we spiral in on the truth.
>
> (Goldacre 2014)

As I understand it there has been a growing academic field of what is called curriculum studies, where those interested and concerned study curriculum theory amongst other things. (In fact I did a part-time M.A. course in curriculum studies in the 1980s.) However, most curriculum theory is about at least two other things: knowledge and learning. Most discussions within 'curriculum theory' also contain a view about the moral context to a curriculum: why should these people, these children, these adults, these nurses and so on, acquire this knowledge, in this place at this time? Is it good for them; is it good for society as a whole? Will it maintain or challenge the status quo, the balance of power in society? Professional education certainly has a moral agenda, and I shall be discussing this later. One key facet of curriculum theory is to ask whether educationalists, institution-bound or not, believe people are to be 'instructed', or 'trained', or 'taught' to enquire for themselves?

In the late 1970s Ernest House amongst other scholars wrote about the clash between craft and technology in approaches to the curriculum. He was alluding to a general social trend at the time, a reaction to a more mechanical, automated, inorganic tendency in the organisation of everyday life. This trend also saw a shift in attitudes in health and social welfare in general which promoted the concept of 'the social model' of the person in regard to an understanding of the contexts to everyday life. One person's illness might well be bound up in a more complex set of social relations of which that person was a part. House et al also wanted to emphasise that changes in thinking about education and the curriculum reflected a more humanities-style orientation: artistic, craft and organic rather than a mechanical, automated and probably reified perspective on life.

These are complex issues that are the 'bread and butter' of curriculum theorists. The normative role of the curriculum is therefore important: what is the social norm of how social workers work, or nurses, and assuming we still all feel that is fine, should it go into the curriculum for the next generation of these practitioners to do the same? However, curriculum theory also contains a critical strand, an argumentative approach that suggests we need to acknowledge who has power to put certain forms of knowledge into a curriculum (perhaps the knowledge that privileges those with most power?) and to be honest about what is happening in these processes. I have already addressed questions about power and authority, and here is another example of that. Whether curriculum designers acknowledge it openly or not, they are making choices about content, about which bodies of knowledge in which subjects or disciplines, should be in and which should be left out. A brief look at the recent-ish history of the national curriculum for schools in our society is a relevant example. Subjects are considered core, some peripheral, some not needed at all. Recent years have witnessed the deliberate marginalising of music and social studies along with other aspects of the school curriculum deemed unnecessary, even a distraction. A certain anti-humanities technical rationality has dominated choices. There are here some interesting comparisons between public and privately funded schools. The latter tending to put a high value on 'the arts' and sport for example.

Education is a practical activity, but to engage with it requires a good deal of philosophising, thinking, conceptualising. In a society like ours we have become used to the formalisation and institutionalisation of education, and whether this helps the dissemination of knowledges, or adequately addresses the complexities of learning, should be discussed. Everyone involved in the process of designing a curriculum where the intention is to transmit bodies of knowledge, skills and techniques needs to ask broader questions about the societal contexts of this process. Why do these people need to learn that? Is this a formative experience, an 'education' that is going to lead to something else, even more education, or is it summative, once done that is it, box ticked. Anyone designing a curriculum will have decided what they wish the outcome to be, for example what sort of person, with what kind of understandings and skills etcetera, are going to emerge at the end of this process?

One aspect of this for the education of professional practitioners that I address in the following chapter is whether it is appropriate to include aspects of social scientific bodies of knowledge into a curriculum. I also ask whether these bodies of knowledge need to be devised, re-contextualised usually, by a professional social scientist, a specialist. Most curriculum theory does emphasise the appropriateness in attempting to broaden the educational experience of learners, and may persuade some course designers that such 'specialists' should play a part. I certainly benefitted for many years from decisions to include me in both design and delivery teams.

A relevant example of this issue from my own experience, from the mid-1980s, was when I was a member of a large course team designing nursing and midwifery degrees. In addition to a broad range of health care professionals, there were a few social scientists, and natural and physical scientists. Initially the scientists were 'responsible' for modules to be delivered at the beginning of the course. These were part of a core curriculum that all students would experience regardless of their eventual health care speciality. Most of the scientists offered a body of knowledge, and appropriate reference to methodologies, which was lifted virtually unaltered from what was on offer to their regular undergraduates in the science subject curriculum. When other members of the course questioned the suitability of this 'recipe' knowledge many of the scientists were surprised. Negotiations took place, and eventually what emerged from the thinking process were modules that reflected the specificity of health care professionals needs. A more problem-oriented approach to those bodies of scientific knowledge was created which had greater relevance to what knowledge would be appropriate for a graduating nurse or midwife. The nurses and midwives in the course team with their extensive experience of practice education made a marked contribution to these discussions.

Curriculum theorists have also acknowledged that it is not just knowledge and skills that are transmitted in the educational process, but attitudes as well. Who is to legislate for this?

Basil Bernstein amongst other curriculum theorists consistently argued that a theory about curriculum comes with a theory of knowledge, which takes us into the realm of epistemology, the philosophical study of the nature and limits of knowledge. Not the least of the issues here is to ask what can people know, and how can they know it. What is this knowledge? Whose is it? How do 'we' acquire it, and does it help 'our' situation? This in turn focuses our attention on learning, and what is it that a person is capable of learning, and how can that be done?

There are many definitions of learning, but essentially it concerns a person's capacity to reflectively draw on their past experience to help understand their current life and consider actions for moving forward, accumulating new knowledge along the way (Abbott 1994).

This definition emphasises the motivation of all people to be engaged, to take action, in their own development. It incorporates issues about cognition, about the experiential and the experimental. Of course the reasons that will take any person over an 'educational' threshold is a key issue, so motivation and relevance are major factors. I remember the stand-out feedback of nursing and midwifery students when they engaged with the degree courses I helped to design: 'When do we get on to the wards rather than spend time in a classroom?' I certainly learnt from that one!

The majority of writers (including myself) on adult learners have emphasised some core expectations, criteria even, written into student programmes. For example a student entering an undergraduate course would be expected to operate judgement over what information would be relevant to a particular task or assignment, and what material they could summarise, edit or even omit. Students would be expected to point to links between ideas and theories across their course. I cite later in this essay an example of the 'ideas bridges' that I would regularly write for students to assist in their orientation from one 'module' of their course to another. It will be obvious to a course designer what ideas, theories and information learnt in one part of a course could form the basis of understanding the next. But this cumulative approach to the acquisition and use of knowledge may not be apparent to all students. Tutors do of course have a responsibility to guide (perhaps quite literally) their students through a programme, but the more independent students can become in making these connections the more they can develop analytical skills and exercise judgements they feel confident about. Students will be regularly tested by their tutor(s) on how effective their learning has been. And, I have always argued a full range of options exist for this basic task. I recall suggesting to my sociology undergrads that if an exam was required by the course regulations they might prefer a seen paper, mull it over for a week, then sit the two- or three-hour test. I was surprised by just how many students demurred and said they would prefer 'a surprise'. Pressure of other coursework seemed to be the main reasoning here! Students in general, and I feel particularly so those doing a professional practice course because of their role in 'the people industries', are expected to develop the skill of argument, both on paper and in gatherings like seminars. The social-interaction dimension to the latter is particular important, and inevitably some students are better at 'public' speaking than others. But both means are a learnt skill, and tutoring students in this should be a high priority at the beginning of any course. At the very least a basic understanding of the dialectical method is essential, and in my experience once students have grasped the thesis/anti-thesis/synthesis idea they appreciate how this helps them to organise the bodies of knowledge, including theory, that they must learn how to manipulate.

Adult students returning to education, whether full or part time, in or close to degree level, need help with re-orientating themselves to study, and to be left in no doubt just how demanding this will be. This is not to discourage potential students – there are plenty of those barriers already – but to be realistic about preparation. If I knew of students who were planning returning to education I would recommend their seeking out and attending a return to study course that would in the right hands assist their transition, help them get back into a studying regime, and improve their confidence.

All responsible tutors in and around adult education hold these basics as essential learning theory, emphasising that a theory of adult learning, and how adults might and do learn, rests on a clear-eyed understanding of such learners.

Discussions about learning in recent years have also highlighted that there are different learning styles, which in addition to categorising 'students' for the benefit of curriculum designers could also help learners to gain an insight into their preferred way to organise their learning. Honey and Mumford (1986) proposed four learning styles:

1 Activists, who involve themselves in new experiences and tackle problems by 'brainstorming'.
2 Reflectors, who like to stand back to ponder experiences and observe from many different perspectives.

3 Theorists, who like to analyse and synthesise and focus on assumptions, principles, theories and models.
4 Pragmatists, who like to try out ideas to see if they work in practice and take the first chance to experiment and apply.

Of course these are only theories, guidelines for more discussion, more research and evidence collecting, but they do have the virtue of reassuring learners, especially so older ones, that there are different ways to engage in the educational process. Who they are as a fully functioning person does not have to be left at the classroom door.

I will return to these issues in more detail in the next chapter as one key aspect of discussing the specificity of professional education and the role of social sciences within that process. One of issues which I raise there, and indeed develop further in my chapter on roles, concerns emotional literacy as an aspect of the learning process.

> Emotional literacy is made up of the ability to understand your emotions, the ability to listen to others and empathise with their emotions, and the ability to express emotions productively. To be emotionally literate is to be able to handle emotions in a way that improves your personal power and improves the quality of life around you. Emotional literacy improves relationships . . . makes cooperative work possible, and facilitates the feeling of community.
>
> (Steiner 1997)

Bold claims, but certainly not without evidence-based justification. An example of this situation from my teaching experience would be the countless number of times I had to support and mentor 'mature' students in the higher education process. Despite the initial motivation, and the courage to actually take the decision and step over the threshold of taking on an educational programme, many older students can easily be overwhelmed by 'returning' to this very institutionalised situation. Taking such action they open a veritable 'Pandora's box' which requires them to take on a role, with all the attendant rules, to work alongside other role holders, and to engage with a diverse range of other specific role performers, their tutors. They could become afraid of missing some crucial link, not responding as they thought they should, feeling an outsider amongst all the younger, apparently more confident students and so on – a potentially long list of hazards to be navigated while possibly devoid of the appropriate map needed. One way I sought to reassure them that they could do it was to draw up a list of their existing life skills, the ones that had served them very well up to this moment of panic. I am pleased to say that the early orientation support given to students on all manner of courses has improved a good deal.

This understanding of engaging with the learning process is even more important for those students embarking on an education for professional practice simply because so many of their clients are going to feel exactly the same. An understanding of empathy and ethics might need to be on the first page? One important aspect of this issue goes beyond what students learn from the curriculum offer, how and why they do this, and crucially why any course designer would wish them to do so. Students in professional practice courses are asked to come to an understanding of considerable bodies of knowledge about the contexts to their roles, and role performance and the nature of practice. Assessment for such students should be emphasising the use-ability of these knowledges that are acquired by whatever teaching process used. Students of professional practice should be deep rather than surface or shallow learners because only an understanding of the core and key principles of these knowledges will enable them to acquire the 'ownership' of this foundation to the process of becoming role performers. So better to do 'less' more fully, the crucial building blocks, than skim across several issues. I recall many years ago being asked to read George Eliot's *Middlemarch*; 'just skim it', the tutor said. I thought, even then, what would be the point of that?

Most curricula on offer will contain what I have called 'recipe' knowledge: A + B, with a bit of this and that will produce outcome C. There it is on the page, pay attention, follow the recipe, and the end result fits the norm. This is usually called the transmission model, and it is still the favourite way of regarding teaching, learning and assessment. The aim is summative from the outset and leaves little opportunity for a creative and imaginative grasp of the problem under consideration. This is a shallow approach, and in my experience students in higher education can make their way to a decent degree while grasping very little (apart from how to pass the tests) and certainly lack the ability to apply what they 'know' to other problems not actually on an assessment schedule.

Deep learning is the opposite of this approach and accentuates the social constructivist notion of teachers and learners being reflexive and reflective, cooperative, argumentative and open-minded about the whole process. Learning is a dialogue, a dialectical process of thesis, anti-thesis and synthesis. Assessment should be problem solving–centred, and collaboration between the general professional standards bodies, institution and teacher taught. Most teachers in higher education still play a major role in devising assessment for the courses that they (help) design and deliver, and discuss with the wide range of stakeholders involved the form assessment could take. In my many years of setting and marking assignments I was aware that many colleagues routinely grumbled about having to do marking. How they could adequately comprehend the progress of their students' learning and understanding without reading, discussing and responding to their work remained a mystery to me.

In my next chapter I cite a 'student-centred' learning project from the late 1990s, but at this point underline the many discussions about such an approach that happened a decade earlier:

> Most traditional courses are described and indeed predicated on what the teacher does. A student-centred course makes a clear commitment not merely to focus on the student activity but may also give students choice in the directions their learning takes. It therefore involves considerable delegation of power by the lecturer and an equivalent assumption of responsibility by the students. The lecturer cannot of course completely deny the authority invested in her/him by the institution but that authority can be clearly delineated in contractual statements about who takes responsibility for what. In a student-centred course the students will typically be presented with clear choices about what it is they want to learn, the order in which they do so and how the learning should be achieved. It may also include a large measure of self and peer assessment. How ready the students may be for this approach and whether they are likely to need induction or training for it is questions that also need to be addressed.
>
> (Jaques, Gibbs, and Rust 1990)

I certainly tried to be as open and democratic as possible in designing and delivering courses. For example, the management of a seminar programme for my various undergraduates was usually given to them to develop in a cooperative way. Of course I had already devised a seminar programme outline (but invariably based on previous student feedback) and provided an indicative reading list. As Jaques and his colleagues suggested, some measure of continuing social control is inevitable. I would ask my students to elect one of their number as coordinator of the schedule – who was going to deliver what seminar paper and when – and always encouraged students to look beyond the suggested reading. After all they did read other books in other subject areas. They might even be bold enough to offer up for discussion ideas from fiction! I might not always be present at all the seminars for all of the time, but was always close by if required. I tried to emphasise that this was 'their space' as well as our space, a collaborative experience with their peers, but much more formal (and often assessed by them or me, or both) than their usual times together in twos, or threes, over coffee, sharing notes, working out an interpretation of what was going on and devising responses.

A further aspect of these course design and process issues is the active researcher role of academic staff in higher education and elsewhere. Most staff in higher education are subject specialists and as such will be engaged in a continuous process of updating and developing their ideas via research within the context of their subject specialism. The nature and form of the research will vary considerably depending on the value orientation and research aims, and methodological style, of the particular person.

Elsewhere in this book I discuss the role of action research in pushing forward the boundaries of knowledge while engaging with a broad range of research subjects, the people who are in one way or another the focus of the research. The very nature of current wisdom of practice, the paradigm perhaps, is affected by debates around the appropriate evidence to support practice interventions. There is for example a lively debate about the differences between an evidence-based practice, which is raised again in a different context in the next chapter, and an evidence-informed practice. The latter concept offers a fairly conventional triangulation of academic research, expertise based on practice skills and knowledge and the experience of users and carers. However, the field is uneven, and despite numerous studies emphasising just how critical a research-driven up-to-date evidence base is for health and social welfare practice, some areas, notably social care is under-researched. Within the last decade there have been a number of state and parliamentary reports critical of the lack of reliable and accessible research on health and social care policy interventions, changes to professional roles and practice and so on, that pass most basic tests of the veracity of the available evidence. One frequent question has been whether there is any significant evidence that measures the outcomes of various policy and practice interventions into health and social care provision. Does doing this or that, or not doing something, actually make any difference to the lives of those in receipt of care and services? As I have said elsewhere in this essay, the knowledge base, formal and informal, upon which daily practice actually rests, and where 'theories' (implicit and explicit) of care are the underpinnings of this practice, is both in regular turmoil and rarely questioned.

Research organisations like Research in Practice in Adult Social Care (RIPA), based at Dartington in Devon, are developing a range of interventions for organisations and individual practitioners aimed at filling these gaps.

> Broadly our work falls into two categories. One, we focus on the building blocks of good practice, developing learning and development resources to enable those working in health and social care to carry out their work effectively. The second area is more "topic" focused, providing learning and development resources to enable professionals to understand and work with specific issues facing adults and families accessing services, such as dementia, self-neglect, and working with people who experience multiple and complex needs.
>
> (www.ripafa.org.uk)

RIPA uses a wide range of means to deliver these resources, which in part could help to counter one of the major problems with creating a more evidence-savvy practice, namely relatively easy access for practitioners of all kinds. Most professionals will argue that they simply do not have the time, or encouragement, even if they had the inclination, to work their way through lengthy and often over-academic-style material.

The scholarship that is a core aspect of academic work does feed into both the curriculum offer any department makes to its students, and thereby the ideas that those students are exposed to. Teaching is one part, even half, of the dual roles that most academic staff in higher education perform. Students are invariably exposed to the development of ideas, research problems and so on that is the everyday work of the researcher/member of staff. The ongoing outcomes from any one person's research will feed into the course on offer. In my experience students may actually be involved directly in the development of that research. The education of professional practitioners is a clear case where this is taking place, especially as so much of what is done concerns problem solving.

In my next chapter on curriculum issues, and most certainly in the following chapter on roles, these issues about theory will be a major focus because so much of what happens in everyday professional practice is bound up with the range and role of theory and the value which is placed upon theory within the regularly changing body of knowledge which underpins practice.

Chapter 3

Curriculum issues

In this chapter I will discuss the 'what' and the 'why' of the curriculum, and present some case studies, a few drawn from my own experience.

Before looking at these issues in more detail it is vital to emphasise that in discussing the creation and delivery of the curriculum, we are entering into the realm of power.

> We can make this more explicit by starting with the assumptions that those in positions of power will attempt to define what is to be taken as knowledge, how accessible to different groups any knowledge is, and what are the accepted relationships between different knowledge areas and between those who access to them and make them available.
>
> (Young 1971 pp. 31–32)

Who has the first and/or final say in the set-up of a course? If a curriculum design team is created within an institutional framework, who decides who is in that team? Who is the chair? Who sits where around the 'table', literally and metaphorically? It is the form that the educational process of designing curricula starts with and takes on that can critically determine what happens next. The value orientation of the participants will play a key role in the way knowledges are stratified, ordered, considered, seen as essential components or not. Decisions are made. Even if as is often the case one person designs, delivers and initially at least, examines course outcomes, that person will have determined what is appropriate to be included in the curriculum. Now this may seem to be patently obvious, common sense even, but a history of cultural judgements has already been made about the relative status of knowledges that will make up the curriculum in question. We are here entering into the theoretical domain of hegemony, the concept that seeks to explain the continued domination of one class or elite group over society through the determining control over ideas, and over what knowledge counts are proper and appropriate in the overall management of society. One form that this takes is through the control of the form and content of educational curricula.

> The business of organising education – creating types of institution, deciding lengths of courses, agreeing of entry and duration – is certainly important. Yet to conduct this business as if it were the distribution of a simple product is wholly misleading. It is not only that the way in which education is organised can be seen to express, consciously and unconsciously, the wider organisation of a culture and a society, so that what has been thought of as simple distribution is in fact an active shaping to particular social ends. It is also that content of education . . . what is thought of as "an education" being in fact a particular selection, a particular set of emphases and omissions.
>
> (Williams 1965 p. 145)

Basil Bernstein often highlighted the issues raised by Williams, and emphasised that the management of education, especially given the hierarchical nature of society couple with a driving ideological set of aims would appear to create a 'closed' form. However, once decisions are opened about the design and development of a new curriculum, all is thrown into 'flux', the outcomes of which can prove to raise many issues that challenge orthodoxy. In the field of health and social welfare education in general this period of flux has created as many questions as clear answers.

The course design 'team' will set the opening parameters for the discourse about form and content that will take place, and as I discuss below, this can be a significant power struggle. Language and identity is one issue here, especially so when people are gathered together from a variety of professional practices and academic fields. So, before looking at a specific case study, I also need to acknowledge some important theoretical and methodological shifts that have been significant.

At the same time as the considerable increase of professional education courses in higher education from the 1960s there was also, as mentioned in my introduction, a significant turn towards critical theory, a more challenging approach to social orthodoxies, and the rise within the welfare professions and the educators of a focus on reflexive, reflective and research-minded thinking and practice. A familiar pattern emerged of seeing aspirations for these approaches as an overlapping of 'critical action' with 'critical reflexivity' and 'critical analysis'. The first of these perspectives included the creation of a sound skill base linked to an understanding of context, of which structural disadvantage and demands for empowerment were an organising focus. The slow but sure impact of feminism and anti-racism was a factor. Secondly the practitioner as self-conscious, and self-questioning emerged from the reflective practice school of continuous professional development. Thirdly the standard approach of critical theory came to bear on all this by emphasising the necessary and central role of an explicit focus on theory in challenging orthodoxy and the status quo. There has always been an important element within critical theory that focused on conscious and unconscious aspects of the person, including discussions on the value of psychoanalysis. This focus on a fuller understanding of the self has certainly underpinned the development in ideas about the 'whole' person.

It was/is often for the first time and via a community action focus (tenants' rights campaigns etc.) linking up the common sense values of local people with a body of theoretically led knowledge illuminating previously hidden realities of everyday life. This expansion of consciousness is clearly a dimension of the ongoing and on-the-ground struggle against the dominant hegemony, challenging the usual recourse to 'there is no alternative' addressed above.

At the centre of this approach is the use of multi-perspectival research methodology, including the reflexively autobiographical, to challenge existing power hierarchies in society in general, and within the professions themselves – all of which contributes to us trying to make sense of ourselves. In my many years of teaching a wide range of students, moments of epiphany were clear to see and celebrate.

A good deal of emphasis has been placed in recent years on the concept of the 'research-minded practitioner' (RMP), and the criteria created by Angela Everitt and colleagues is a good example. They argue that the RMP will:

1 Be constantly defining and making explicit their objectives and hypotheses.
2 Treat their explanations of the social world as hypotheses – that is, as tentative and open to be tested against evidence.
3 Be aware of their expertise and knowledge and that of others.
4 Bring to the fore theories that help make sense of social need, resources and assist in decision making with regard to strategies.
5 Be thoughtful, reflecting on data and theory and contributing to their development and refinement.
6 Scrutinise and be analytical of available data and information.
7 Be mindful of the pervasiveness of ideology and values in the way we see and understand the world (Everitt 1992 p. 5/6).

One dimension of these issues mentioned above that has now entered into the core of teaching on reflexivity and reflective practice in professional practitioners is the idea of critical practice. As discussed above, critical practice seeks to draw together the dimensions of reflexivity, i.e. the engaged self, with analysis, the active use of knowledges, and action, and then using all the aspects cited above to work on empowerment. This was and still is usually seen with regard to the client (and client group) but I want to expand this goal to practitioners themselves.

I would not underestimate the initial difficulty in making sense of this approach, but the long-term benefits of having these analytical skills in our 'intellectual toolbox' makes the effort worthwhile.

Item 2 of Everitt's list is crucial in this respect because one sure way to empower both clients and practitioners is to encourage them to ask 'why?'

or even 'why not?' I will consider some the implications of these issues later in this essay while thinking about values and roles. One key aspect of this critical approach is the core cultural sense of fairness that resides in our society.

The reader might also wish to see my chapter 'Turning the concept of reflective practice into curriculum reality' in Astley (2006). In this already cited book on professions and practice I question the usual shallowness of the application of the concept of reflective practice. Since this concept became *de rigueur* some decades ago no professional education curriculum has been without its Tuesday afternoon slot where the box for having 'done' this concept is well and truly ticked. It is commonplace now for professional practice courses to contain a whole module dedicated to reflective practice, and an understanding of the concept-in-action has become a testable standard. It has also developed some thinking-in-action methodologies like the reflective journal. This task is not to be underestimated given the opportunity here to create a thinking space while writing.

A further influence on setting priorities for curriculum development has been seen in recent decades with the rise of a movement for evidence-based practice in professional practice, particularly so in the provision of health and social care. In some instances Local Authority Social Services departments contributed to the development of this approach, and sometimes with university-based social work education initiatives as with Brian Sheldon in Exeter in the 1990s. I will be saying more about this later.

It is often media-driven reporting on safeguarding issues and the like that brings the health and social welfare 'community' to revisit the fundamental issue of professional standards. When a major media coverage safeguarding, or even general governance event happens, even the state will enter into the fray with an inquiry and report. This was true in 2003 with the government publication of 'Every Child Matters' report and subsequent Children Act in 2004 which set out what were considered to be key safeguarding criteria. This then set both the tone and statutory requirements for the work of local authorities in all these respects. The current educators of professional practitioners freely acknowledge the way in which this particular set of safeguarding obligations and tasks has changed the nature of practice and made the task of recruitment of social workers to local authorities much more difficult. Education providers also comment on the decrease in opportunities for local authority student placements, and that students are increasingly wary of taking up posts in services for children on graduation. As I discuss in this essay, the increase in alternative providers of health and social welfare has been matched by the desire of graduates in this general field to look beyond local authorities for a post-graduation role.

Of course the maintenance and policing of appropriate standards is always on the agenda of health and social wealth providers and those who devise and provide courses for professional practice. Indeed perhaps the key aim of this educational is to develop graduates who have 'fitness to practice'. The professional associations like the British Association of Social Workers (BASW) regularly set up inquiries into particular, and again usually media-sensitive, issues that arise. Ethics, safeguarding and good governance are always on someone's agenda.

Education providers are made acutely aware of appropriate standards by regulatory bodies like the Health & Care Professions Council that sets out 'duties' for education providers. Amongst other criteria there is 'the curriculum must remain relevant to current practice', 'integration of theory and practice must be central to the programme', 'the delivery of the programme must support and develop autonomous and reflective thinking', and 'the delivery of the programme must support and develop evidence-based practice'.

These criteria clearly return us to issues about experience-based bodies of knowledge. However, paradox abounds in the notion of the 'autonomous' practitioner as many, if not most professionals whose everyday practice is tightly controlled by statutory and management rules, would probably settle for semi-autonomous! This can be yet another dimension of the contradictions found in the set of values held by the graduating professional. Values that are held and developed pre-, during and post-graduation can make them look outside the state apparatus for employment and self-actualisation.

As briefly discussed above, a number of UK university departments and other research focused organisations have devised courses that place an emphasis on evidence-based practice, for example the M.Sc. course 'Evidence based interventions and policy evaluation' at the Department of Social Policy and Intervention' at Oxford University. The course information sets out a familiar approach:

> In all societies, policies are developed and professionals from a range of disciplines and backgrounds intervene to ameliorate social problems in areas as diverse as mental health, child protection and support for parents, education, social security, housing, refugees, care for increasing numbers of vulnerable elders, substance misusers, delinquent young people or those affected by HIV and AIDS. The practical imperative of ensuring effective use of finite resources, together with an ethical imperative to demonstrate that intervention is doing more good than harm, require that practice be based on sound evidence.
>
> The course in Evidence-Based Social Intervention and Policy Evaluation emphasises research methods for evaluating interventions and policies, as well as the advanced study of evidence-based practice.
>
> (From the departmental course site)

As promised above, I now wish to discuss an example of the varied contributions that go towards the making of a curriculum at a particular time and in a particular place. In this instance it was in Oxford in the mid-1980s.

This case study raises important questions about the nature and value of knowledge, a ubiquitous theme in this essay. Social scientists as co-contributors are dedicated to the creation of new, emergent knowledge, alongside an examination and critique of residual knowledge. Arguments about whose knowledge is most important, even the audacity to ask that question, reflect the shifts in power that take place amongst and between key players. For example, the balance of power between different practice groups within any professional practice domain has changed. This case study is an example of this when thirty or so years ago nurses and midwives gained more control over what constituted the valuable knowledge that should make up the curriculum for the training of those practices. Nurses and midwives amongst others sought to prove that they like medics had a theory of practice in addition to, developed from, the bodies of knowledge derived and reflected upon from practice experience. From the 'outside' most specific professional practice can seem both traditional and chaotic. So how to make sense of it, draw some coherent understanding of what determines both form and content of that practice? It was this 'big picture' that we sought to comprehend.

It was also the case that some natural and physical scientists lodged in the academy felt that their bodies of theory should be privileged. Health and social welfare along with some other professional practitioners tended to disagree, and a curriculum content compromise was sought. From the outset curriculum team meetings were characterised by discussions, and arguments, about the relative status of knowledges that would form the course. There was much consternation in certain quarters over the challenges to using existing course materials, for example drawn from the natural and physical sciences. If this course material was good enough for undergraduates studying chemistry, why not nurses and midwives?

Bound up in such responses is the usual issue of knowledge status, where many 'traditional' academic disciplines assume a privileged position in the knowledge hierarchy. (We now see this every day in the school curriculum.) I would also add that in many years of teaching in higher education my experience was that most full-time academic staff tried to avoid doing 'servicing', i.e. providing a small subject specialist slot of teaching in a hybrid practice course. Leave that to the part-timers was the usual response! But this often meant that many academics had little or no contact with health and social welfare practice or practitioners. In the recent years the budget cuts in H.E. (higher education) has usually meant even these 'external' subject specialist contributions have ceased to exist.

So, given the situations highlighted above in the nursing/midwifery courses design process a radical break had to be, and was, made via the suggestion that we, as the course team, should ask those currently practicing 'on the wards' what they would expect a graduate from this degree process to know and be able to apply to their understanding of everyday practice. We were duly informed, which set all course design team members to think very critically about what exactly would be needed to create a curriculum fit for purpose. Perceptions of role making, taking and development were changed in that moment.

In my experience the role of most social scientists engaged in these curriculum debates encouraged a more open and democratic decision-making process, and regularly challenged orthodoxies about whose knowledge was most valuable. It should also be added that there were only a handful of participants in that team who had had any recent engagement with the history and development of curriculum theory as a 'separate' and influential thinking tool for this task. I would also argue that a further issue here is the dual roles reality for most higher education staff: academic/subject specialist on the one hand and teacher on the other; many hours devoted to keeping up to date with the minutiae of developments in the former, while remaining blissfully unaware of the latter.

The year-long process of designing the nursing/midwifery degrees had a profound influence on my continuing 'work in progress' thinking about curriculum design and delivery. I took away from this process several important lessons about how the discourse around the ranking and deployment of knowledge for and in a curriculum required reflection in action. I came to understand that the current form of the curriculum at any time would invariably mean that something else was missing from it. What was it? Who should I consult to enlighten me? What had I, and colleagues, overlooked from the historical debates on form and content? At the very least the narrative of our collective identity needed examining. What was the dominant discourse of that year? What was our vocabulary of motives?

For example, were practitioners to be encouraged to see themselves as change agents engaged in a transformational set of processes that would liberate themselves, clients and perhaps even society? This creative practice would challenge the conventional wisdoms of the day, open up the quotidian for examination; the 'explanation of the social world as hypothesis' idea was taken from critical theory, and the use of research and theorising to argue for a new and more democratic and appropriate reality. One of the implicit, and often explicit, aims of these approaches was to recognise that critical theorising, and radicalising and democratic practice would lead to praxis: a focus on social action for change. There was recognition here that a profession not only has a practice, but also a theory of action in which that practice can become a reproducible, teachable and valid technique. This means that the job of professional education consists not only in the teaching of technique – delivering a baby, schooling nine-year-olds, or working with homeless young people, say – but in teaching the methods where behavioural worlds can be created in which these practice techniques can actually work. This emphasises again that practitioners are not just individuals; they are also members of culture groups who have a 'structure of feeling' (a framework of practice know-how) and an aspect of particular organisational arrangements, custom and practice. These 'setting' features invariably contribute to a context of 'theories in use', which is often in contrast to the 'espoused theories' that professional groups promote into the public domain beyond their group 'boundary'.

The transformation of the social world, a context of practice, has to be an aim of raising up and changing practice itself. One of the outcomes of these approaches is to challenge the tendency amongst professionals' and practitioners' claims that this or that theory of action guides practice. However, these espoused theories are invariably not the theories in action that do actually guide practice. Organisations and their practitioners often claim to be certain kinds of people, to have values of this or that nature, but in fact rely upon something quite different in their everyday practice. They say one thing and do something else! And the something else is often the daily reproduction of hierarchies, power inequalities and undemocratic practices – maintaining a status quo of unfairness perhaps. The routinised and deferential behaviour of practitioners 'encourages' the same amongst their clients. A fatalism sets in. 'I'm only doing my job', 'it's the rules', 'my hands are tied' and so on.

So, in discussing the 'what' of the curriculum, what is included? Where are the key, inevitably historical, issues dear to the hearts of social scientists, and also social science as embraced by non-social science specialists, to be located? Even the nature of content is a major issue to address because of what I have said above about the personnel assembled to argue over and determine content. In an echo of the case study cited above, many of the content battles that I encountered in the 1980/90s concerned the belief among some social scientists that what was good enough for their subject specialist discipline undergraduates was also suitable for the education of professional practitioners. I argued then, and now, that this is not necessarily the case, and that for example, any curriculum designer might be advised to work 'backwards' from the usual perceptions of everyday life of a particular practice domain towards what might be a starting point to devise an education. A detailed examination of roles might be such a starting place. I am not advocating a 'watered-down' version of a conventional social science undergraduate curriculum, but a focused selection of key conceptual and analytical approaches best suited to the practice under discussion, plus a questioning critical edge to challenge orthodoxies. This should also embrace a full and frank discussion about evidence collecting methods and what happens to that evidence over time. Therefore a key factor in opening up a curriculum, old and new, to discussion will sooner or later face the question of exactly how much agency do all the stakeholders in this educational process actually have? Far too many 'radical' re-writings of curricula have not actually addressed the fundamental issue of clients'/users' agency. For all the enthusiasm of educators to promote change in the curriculum, most have been 'top-down' solutions, where the educators have continued to exercise social control. Reading lists for new curricula are usually an interesting place to start in an examination of just how liberating these changes are.

But, and this is a big but, I wish to focus on 'why' questions as well as the 'what' ones. Why did those responsible for the education of new generations of professional practitioners believe that social science should form a core part of the curriculum? Why did social scientists in their many guises believe in this approach to the extent of devoting much thought, time and energy to this project? Why have more recent generations of professional practitioners, turned into course designers and teachers and recipients in their time of these bodies of social scientific knowledge, continued to deliver from this convention?

As a key aspect of their value orientation, was this venture mainly enlightened self-interest on their individual and collective parts, or was there a fundamental commitment to the value of social scientific methods, bodies of knowledge and understanding that the education of 'people workers' would be deficient without these interventions? Without this focus on key concepts like 'role' it is difficult to move on conceptually.

The concept of andragogy was addressed in my introduction and it is important to say more here about adult education and the curriculum. In 1973 the American educationalist Malcolm Knowles, with a background in humanistic psychology, published his book, *The Adult Learner: A neglected species*, which became a key text for those on both sides of the Atlantic concerned with the lack of theoretical speculation concerning adults in the education systems of society. Note that the title of Knowles' text is 'learner' and not simply 'student'. Knowles sought to emphasise the need to think about the special nature of adults involved in an educational process, and therefore the value of theory to acquiring a deeper understanding about how adults learn. Teachers of all kinds should go beyond the primary socialisation and social control–focused curriculum aimed at children to reconsider the world of adults engaged with learning.

> A good theory should provide both explanations of phenomena and guidelines for action. But theories about human behaviour also carry with them assumptions about human nature, the purpose of education, and desirable values. The better you understand the various theories, therefore, the better decisions you will be able to make regarding learning experiences that will achieve the ends you wish to achieve.
>
> (Knowles 1973 p. 2)

The 'you' in question might be either a teacher or a learner.

Knowles' theory of adult learning emphasised the motivation of people to know more about their world and themselves. He also stressed that the past experiences of potential and actual students, with all the inevitable ups and downs of life, should play a central role in their educational futures.

Knowles also sought to discuss the way in which 'situations' related to an adult's life course usually determined entry, or re-entry into education. This is one reason why teachers in higher education consistently comment on the extra commitment shown by 'mature' students to the demands and value of learning (even their specific subject learning). Situations in their lives have brought them to make choices that are acknowledged as significant in their development. Entry into courses for professional education is replete with such adults, who will of course bring their experience of life, and work, and relationships with them. Knowles and the many educationalists who have followed in his speculative footsteps insist that these factors make 'adult education' worthy of special consideration. This is certainly true for both curriculum designers and for those who deliver courses. A common criticism of Knowles' approach has been his clear focus on the individual (even in discussing 'situations') and not enough consideration of broader social and cultural contexts.

My own engagement with social work students would start by suggesting that 'we' should start by exploring the nature of the social upon which we, and most certainly they, would be working. What do we know about it? What do we think we know about people, their lives, their situations, their highs and lows? Are these people the same as us or different, and how, and why, and with what consequences for them, and us?

One key writer following Knowles is Peter Jarvis, whose 1983 book *Professional Education* expanded our understanding of why adult education is different within learning theory, and sought to direct that discussion at the education of professional practitioners. One important point that Jarvis makes is that given the increasing division of labour within the professions the specialist bodies of knowledge have increased, which in turn raises questions about the extent of learning that needs to take place. This is especially true for newcomers to a profession, but does of course raise questions for continuing professional development in general. I raise later an important aspect of this issue when discussing via a case study example the recent tendency to reconfigure health and social welfare provision to use more generic practice roles. So choices about what is in and what is omitted could be a course design and delivery factor, but also a policy maker's approach to reforming provision. Who makes that choice from the range of potential stakeholders: education managers, frontline teachers, students, professional bodies, pressure groups for particular client groups, even clients themselves? Not the least of the issues here is the increasing amount of legislation that is part of these bodies of knowledge. As more research is done on professional practice, and with the continuous feedback from those engaged in face-to-face work with clients, experience is turned into recordable and repeatable bodies of knowledge. As discussed this could be coupled with another demand on practitioners with the increased enthusiasm for evidence-based practice. So practitioners of all kinds are encouraged and required to acquire and accumulate more bodies of knowledge. What balance is struck here with encouraging practitioners, including their 'teachers' where appropriate, to spend equal amounts of time to come to a fuller understanding of the value and relevance of what they already know? This is very much part of the social value of knowledge, especially so in its application to social problems of whatever kind. This situation over several years could certainly have contributed to calls for the length and status inflation of courses, certificate to diploma to degree.

Of course any course designer has to sift through the available, and sanctified, bodies of knowledge to decide what should be in the curriculum, and as a consequence as suggested above, what should be left out. I suspect that many adults entering a professional practice course will be daunted with the sheer volume of knowledge of all forms that has to be 'learnt'. They may well question, privately or, if they are invited to do so and brave enough, publically question the selection of knowledges in the course content. This will most likely be based on their life experience and common sense response to what is on the agenda. Their life experience may well include being the same kind of client they are now seeking to help. One key task for any teacher in response to learner critique or not is to guide students through the content.

This does then return us to basic questions about the aims of professional education, and Jarvis makes a good point here:

> First, the process should produce recruits to the profession that have a professional ideology, especially in relation to understanding good practice and service. Secondly, that the educational process should provide the new recruit with sufficient knowledge and skills, or the continuing practitioners with enhanced knowledge and skills, to enter, or to continue in, the profession. Finally, that the process should result in the practitioner developing an increased sense of critical awareness.
>
> (Jarvis 1983 p. 48)

Alongside of all these criteria sits the key issue of competence and fitness to practice. Sound preparation via a well thought-through education can help with both goals, but post-qualification continuous professional development needs to keep all practitioners up to date with research, policy reports, including those from client-based pressure groups, and shared good practice. Students preparing for a professional role need to have a clear-sighted understanding that their engagement with appropriate knowledges is a long-term commitment.

The case study that follows is an interesting example of the relationships between course choices, course provider's aims and the opportunities available to 'adult' students seeking a career. It also emphasises that the issue of appropriate knowledges to be acquired on a course and on placements can be complicated due to changing contexts, both personal and organisational.

***

Case study

Youth and community work course, Ruskin College, Oxford

When the course coordinator post at Ruskin College came up it felt like the job was designed with me in mind, a college that promotes higher education for adult 'returners' just like me!

I left school at fifteen with no formal qualifications and spent many years studying part time and in the evenings and weekends making up for the deficit in my education and thoroughly enjoying the experience of learning.

As a practicing youth and community worker and youth work trainer, I had been a member of the working group that developed the Foundation Degree in Youth and Community work and came into the post only six months into the start of the first cohort.

I went on to work at Ruskin College for five years variously as course leader, tutor and placement coordinator for the foundation degree and later the BA hons in youth and community work.

My professional life as a youth and community worker mirrored my academic one in that I had worked my way up through the ranks as a volunteer, through part-time paid youth work, full-time training to obtain my professional qualification, full-time work and finally a training role.

On its website, Ruskin College's states its vision of 'a society where everyone has access to quality education regardless of their background, and the opportunity to fulfil their potential' and its aim of providing educational opportunities for those who may be excluded or disadvantaged reflected the transformative potential of youth and community work in enabling individual, groups and communities to take an active part in their own lives, though its tools were not formal qualifications, but a process of informal education.

Unfortunately, my experience of youth and community work in practice was that its central educational practice of informal education was both poorly understood and indifferently delivered at best, with professional training for the part-time and volunteers who made up the vast majority of the workforce focused on issues, such as anti-social behaviour rather than pedagogy.

In consequence, staff found it hard to name the values at the core of their practice, let alone champion and defend it in the face of rapidly changing priorities and funding pressures.

There was a distinct unevenness of quality 'on the ground' with a tendency for youth and community workers to rely on charisma rather than competency firmly rooted in the principles and values of the profession. This 'cult of personality' required no other justification than 'the magic happens because they are with me'.

Youth workers, therefore, gained a reputation, sometimes deserved, for being disorganised, lazy and hazy about professional boundaries. However, some of the criticism came about because workers were attempting to give young people a voice and redress the power and control deficit, something not always appreciated by the institutions in which this work took place.

I developed the course at Ruskin College as a 'safe space' modelling the practice of youth and community work by helping students explore and debate the physical, personal and ethical limits of their profession from a strong theoretical understanding of structural inequality, power and empowerment.

In doing so, I worked hard not to replicate my own experience of professional training. My training course was constructed around identity-based group work, but did not establish this safe space in which the important conversations could take place and instead quickly developed into a hierarchy of oppressions in a way that closed down conversations and established barriers which became insurmountable and defensive. The bottom of this pecking order was inevitably the white men: a deliberate and punitive strategy. I remember one man desperately constructing a positive identity for survival by becoming both Irish and gay!

At Ruskin, I think we mainly succeeded in producing workers who understood and could articulate why and how to practice and whose agenda they were representing.

It helped that our students were drawn from a wide range of ages and ethnic backgrounds, though we always had more women than men to a ratio of 2:1. Many students were parents and carers, bringing much valued life experience to counteract their daily struggle to juggle! Our only entry requirement was experience of voluntary or paid youth and community work and the vast majority had few formal qualifications and were anxious about their capacity to manage a degree-level course. Between 20% and 25% of students were diagnosed with dyslexia, which for some was a huge relief and an explanation as to why their school experience had been so unsatisfactory. However, others found their diagnosis devastating as they understood this to be the start, not the conclusion of what had already been a difficult journey.

That many succeeded was due to their dogged commitment and a comprehensive package of study support including a range of diagnostic tests and ongoing study skills training. Our annual intake was between twelve and fourteen students and we used a weekly tutorial system, like our more privileged cousins at Oxford University where students were expected to present work in progress in groups of two or three, so nowhere to hide!

Over half the credits for the programme were awarded through work-based block placements with a wide range of statutory and voluntary organisations. Organising and troubleshooting these placements was a time-consuming, sometimes frustrating but invigorating part of the job. Most students enjoyed this aspect of the course best as it was where theory and reality collided in a meaningful way. We had some brilliant fieldwork supervisors who were supported by a three-day training course covering practice supervision.

Most graduates went on to local jobs in the statutory and voluntary sectors.

However, in my final year, when cuts to local authority funding was started to bite and the number of work-based students fell, the quality and range of experience that students brought with them also started to decline. Choosing to study youth and community work became less of a step towards a positive career, but something that you did when you didn't have the qualifications and experience to get into any other professional degree-level course, such as social work.

Oxfordshire County Council 'deleted' its youth and community service over seven years ago, two years after I finished working at Ruskin College, so most of the recognisably youth work jobs are to be found within faith-based settings which have their own training providers. However, the Ruskin course soldiers on as only one of two still running in the south-east of England and is still validated by the NYA (National Youth Agency), itself an organisation much diminished.

It is much harder now for these students to find work placements, so I still take one or two a year though I no longer work in a youth and community setting. These students are younger, less experienced and often quickly disillusioned about their choice to study youth and community work, particularly at such high cost with less chance of a job at the end.

Rachel Hills 2/18

***

What are appropriate knowledges is an issue discussed in many parts of this essay, and the theory/practice relationship is a key area for debate. For example, it has always been commonplace for often indigestible amounts of theory to be passed on the students without qualification. It may well be that a range of theories are on offer, and in compare and contrast ways these theories are examined for their value towards explanation and understanding. However, in my experience too many students are expected to digest too much theory without question because of the special status given to this 'science'-type knowledge. The ongoing terror of the 'T word' has been a constant factor in the educational experience of students of all ages engaging with professional cultures. However, my own experience and that of many others, like Keith Brown at Bournemouth University, is that older (so-called mature) students entering this process can really struggle with the jargon, become bamboozled by the course regulations, sucked into the ritualisation of role performance, and very anxious about taking on the specified academic writing style. A good deal of straightforward advice, worked examples and mentoring is required, and I know that this is now a very high priority for course providers.

This also takes us into an area that is discussed in several different contexts in the essay concerns the notion of 'critical thinking', the cognitive twin of reflective practice. My understanding is that enthusiasts for this believe that critical thinking is very much about human creativity, for example about the role and use of language, about talking, for example describing an experience, showing ideas, discussion, argument. It is also of the senses in that it draws on human characteristics like observation. Critical thinking is essentially both a personal activity, the cognitive realm, and has a social dimension because of the fundamental need to cooperate through communication. As I discuss in this essay the concept of theorising as an immersive human experience is very much part, at least an opportunity, to make sense of ourselves. However, despite what individual practitioners may be thinking and how they may be taking action, there are the organisational contexts and professional culture to navigate. Stephen Brookfield emphasised the problems here in his 1987 book *Developing Critical Thinkers*:

> Theories in use are kept private, since they frequently contradict many of the apparently revered tenets of espoused theories. Even when espoused theories do not work, there may be a reluctance on the part of many professionals to make public their critical analysis, since to do so is to appear to be either incompetent (unable to apply theories correctly to specific situations) or heretical.
>
> (Brookfield 1987 p. 153: there is also a longer discussion of Brookfield and his concerns in Astley 2006 p. 58/59)

Some psychologists would also refer to this phenomenon as cognitive dissonance, the mental discomfort experienced by a person who simultaneously holds two or more contradictory beliefs, ideas or values. Either way, there is a lot of it around.

Therefore one of my arguments in this essay concerning 'making sense of ourselves' is more time should be spent on theorising, an aspect of the thinking/reflection-in-action, rather than the inculcation of extant theories.

Given the above, it is important to emphasise that those designing and delivering curriculum for professionals need to recognise the nature of their student population and act accordingly. Many learning theorists, and Kolb's Experiential Learning of 1984 comes to mind, have emphasised the cyclical nature of learning where the learning passes through stages of exposure to 'new' bodies of knowledge, reflection, consideration of use value, action on this information and assessment. What also has to be noted is the way in which any new bodies of knowledge are presented and received. The 'classroom' is a laboratory where social interactions happen in an experiential and experimental way. The information being offered by teachers and learners will/should come in many forms, and thinking through and talking has to be a high priority. Society as communication is a relevant question to pose. How any information is received and the effect of that on any one learner is a crucial aspect of the educational process, and this is one reason why reflective journals as mentioned earlier (in whatever form they may take) are an essential tool for the learner.

I have always advised students to use a 'journal' when working on an essay. Keep that journal on the person at all times and as ideas pop up, come to mind, or are sparked by an event, a conversation, write it down. I recall the aghast look on a group of students' faces when I admitted always keeping a notebook handy when ironing clothes, as we all do! My excuse was that some of my best ideas came to me while in the 'dead space' of ironing. I have always advocated this ongoing thinking and note-taking method for students when writing assignments, keeping the question and relevant issues constantly 'on the move' with cumulative evidence collecting and analysis.

Having raised the issue of the 'classroom' above I should also briefly discuss the concept of the practicum.

> A practicum is a setting designed for the task of learning a practice. In a context that approximates a practice world, students learn by doing, although their doing usually falls short of real-world work . . . When a student enters a practicum, she is presented, explicitly or implicitly, with certain fundamental tasks. She must learn to recognise competent practice. She must build an image of it, and appreciation of where she stands in relation to it, and a map of the path by which she can get from where she is to where she wants to be. She must come to terms with the claims implicit in the practicum; that a practice exists, worth learning, learnable by her, and represented in its essential features by the practicum. She must learn the "practice of the practicum" – its tools, methods, projects and possibilities – and assimilate to it her emerging image of how she can best learn what she wants to learn.
>
> (Schon 1987 p. 37/38)

Donald Schon, along with his associate Chris Argyris, were the leading lights in the discussion on reflective practice in the 1970s and 80s. I discuss reflective practice a number of times in this essay, and the 'learning how to learn' theme throughout, along with also considering the educational process issues associated with practice placements for students.

Schon also made the key point that a practicum-type experience for a student is an aspect of the reflection-in-action that is central to a budding professional making sense of the culture, and cultural issues, that they are entering, taking on. The tutor, course- or placement-based, engages the student in a dialogue, a reciprocal reflection-in-action, that should enable the student to come to an understanding of what is going on in the real world of everyday practice.

But there is something else here as well that highlights the ecology and the environment of student learning. The work of psychologist James Gibson has always interested me in this respect. He coined the concept 'affordance' in the 1960s:

> The affordances of the environment are what it offers the animal, what it provides or furnishes, either for good or ill . . . I mean by it something that refers to both the environment and the animal in a way that no existing term does. It implies the complementarity of the animal and the environment.
>
> (Gibson 1979 p. 127)

It is the notion of 'complementarity' that particularly appeals to my sense of the reciprocal nature of practice education in, for example, the practicum, wherever that might be located. Gibson's work was primarily focused on the socialisation experience of infants and children, but the concept of affordance seems valid to me if placed into the context of what course tutors and placement supervisors, later mentors, colleagues and so on, both set up in advance for the experience of the student, and develop in the educational process. I also like the link between Gibson's ideas about the technological manipulation of an environment as affordances and Schon's concerns with the 'tools' available in a practicum. As I have said earlier the notion of craft as technique within educational theory implies an artisan approach to educational experiences.

There is one further psychological perspective to mention here which compliments what I have said above, and is from Reuven Feuerstein's social interactionist theory of learning, which looks to explain

> the way in which stimuli emitted by the environment are transferred by a "mediating" agent, usually a parent, sibling or other care giver. This mediated agent, guided by his intentions, culture, and emotional investment, selects and organises the world of stimuli for the child. . . . Through this process of mediation, the cognitive structure of the child is affected.
>
> (Feuerstein, Hoffman, and Miller 1980 p. 16)

Three years ago I taught a group of teaching assistants engaged in a part-time learning and teaching degree course. Inevitably we reached a point early in the course where learning and child development was to be discussed. In order to help prepare them for this I wrote the following 'bridging piece'.

***

Thinking about development

When we start thinking about the development of human beings there are, inevitably, many issues to consider, and not the least of these is 'how do we get to be who we are?'

Social scientists in general use the concept of socialisation to describe the process whereby we learn how to be a social being, to take our place in society. Socialisation and bouts of intensive re-socialisation occur throughout our lives, for example becoming a parent, or a student! Of course there is always a particular interest in the development of infants and children, a crucial period in the formation of the self, and the shaping of identity. We should acknowledge that biology plays a crucial part in early development, with the acquisition of motor skills and the physical capacity to walk, speak and so on. There have been many theories offered to explain the processes of learning to be a human being, including whether infants start out as a blank sheet upon which 'society' is written. Usually theorists agree that as the infant and then child develops they take on the culture, the ways of life, language and values of those most immediate to them. This usually means parent(s), family, community, social class, belief group and so on. These key people are seen as agents of socialisation in that they have a key role in the learning processes. However, these key people are also agents of social control in that they seek, self-consciously, knowingly or otherwise, to get us to conform to their way of life, how these agents understand the roles and rules of the host culture. They seek to 'help' us to fit in. As social institutions schools are central to this process, they are essentially 'culture repeaters'. Most children see these agents, these significantly close and important people as their reference group, the people they look to for an understanding of life, to learn life skills, and to love, care for and protect them.

We know that as children age into their adolescent and teenage years their reference groups often change. Young people seek, and become more independent, seeking ideas, advice and guidance from their peers, people of their own age and circumstances. Their 'society' changes, values are re-examined, and new allegiances formed. Of course this is still part of the socialisation process, but the focus of social control changes. It is at this stage of development that most young people begin to see themselves as their own change agents. This can create tensions with those earlier socialisers! Some renegotiation of rights and roles may need to take place to create peaceful co-existence.

One vital part of the development of human beings which is closely associated with values is the issue of empathy. Put simply, this asks whether we are capable of putting ourselves 'into someone else's shoes'. This then in turn poses questions about ethics and our understanding of the moral world we are part of.

Discussions about ethics begin with the 'golden rule' – do unto others what you would want for yourself – and being a sociable human being requires us to deal with this.

I should also mention that there are many different theories about how children learn; and quite often psychologists, socio-psychologists and sociologists do not agree with each other. It is also the case that there are major differences of opinion between philosophers, usually between those called moral philosophers and biologically focused natural scientists. Some of these scientists describe themselves as materialists and argue that we are nothing more than the sum of our biology. Their argument is that we are essentially a human machine: we are wired up to function in certain ways, and how we are and what we do is a consequence of this. There is not separate mind, or consciousness, or soul; our brain is the biological equivalent of the most advanced computer we could comprehend. There are some theorists who argue that we are essentially animals, very similar in our habits to all other mammals.

However, many other 'thinkers' and theorists, like those moral philosophers, and most social scientists, disagree. In their many diverse ways they argue that as we develop from infancy into adulthood we expand our consciousness, our intellect and interpretation of life. These thinkers argue that our mind interacts with the environment in which we exist – physical, social and cultural. So, not only is it true that 'we are, therefore we think', but that 'we think, therefore we are!'

Most of the theorists you will encounter in this course are usually referred to as 'social learning theorists', who argue that we are not 'pre-programmed' as human beings, but learn within and from the social world of which we are a part. As I have said above, there is though a considerable range of ideas on exactly how that happens. I should also mention that the disputed notion of 'free will' can also play a part in these discussions. Many religions regularly contribute to the debates about who we are, what we have to start out in life, and how we develop. While acknowledging the cultural influence of religion, most social scientists take a secular view of development and disagree with those religious ideas that suggest that we are directed by some other-worldly being(s).

So, what we have to consider and discuss are the many different ideas and theories that seek to explain what we are, and who we are as human beings, and to assess the possible, and probable, ways in which we develop. Of course these ideas are important and raise many questions about the human condition. One important aspect of this discussion is therefore called the 'nature and/or nurture debate', and we will look at this in the module. What we personally believe about all this and consider to be the right answer is an aspect of our values, and will influence the way we think about children and our everyday practice as educators. We all have 'theories' about these issues, but may never have been required to set them out, explain them or discuss why we believe this or that. Often the ideas that we have about human beings and their development are never challenged or exposed to alternative ideas, which may or may not require us to reconsider what we think and do. Being an undergraduate requires you to think about these issues and where you stand in these debates, and to come to a better understanding of what exactly informs your practice, and that of your colleagues. (John Astley 11/14)

***

My students told me that these pieces were helpful.

However, as always we do need to remind ourselves that available technologies in the educational domain, as elsewhere, are not power-neutral. Who controls technologies, residual and emergent, is a crucial issue; IT (information technology) is a wonder, but not necessarily a total good if you are a practitioner sitting at a screen all day in a non-interactive relationship with the alleged beneficiaries of their good grace.

Before moving on from the issue of teaching environments I wish to mention an example from my own experience.

When doing my teaching practice in a secondary school at the beginning of the 1970s I learnt a great deal about the nature of learners, and the severe limits to 'teaching'. I had a CSE (certificate of secondary education, offered to non-GCE [general certificate of education] students) class doing social studies, which including some very basic anthropological concepts and examples. One fifteen-year-old, a big lad, was very quiet and did not do much work. At the end of a class I asked one of the other boys whether Brian was always like this and was told yes of course, he is a farmer's boy. That seemed enough explanation. I later said to Brian that I understood he was in a farming family, and did he have any particular interests or jobs on the farm? Yes he did, he bred and managed ferrets that the farm needed. So, being a naive young-ish teacher I asked him if he could bring some of his ferrets into class and talk about what he did. He did just that, giving a front-of-the-class thirty-minute talk and demonstration with two ferrets, talking about the characteristics of the animals, their role at the farm, his breeding regime and so on. The other pupils were completely focused on this presentation and gave a spontaneous round of applause at the end. Brian's behaviour in class changed a great deal after that; he was seen by his peers in a different way and became much more engaged and collaborative. When word of this event reached the senior-year teacher I received a serious rebuke and told never to do this sort of thing again; but I did of course.

One final issue to address in this chapter concerns student-centred learning (SCL), a concept that has been in discussion for many years, but in the UK became a major talking point in higher education in the late 1990s. SCL is also a spin-off from self-directed learning, a key concept in adult education. Student numbers in higher education were increasing rapidly, especially so in the old polytechnics/new universities, while staff numbers remained the same. All resources were being stretched and the state apparatus realised that some internal educational adjustments needed to be made and fell upon student-centred learning. Get the students to do more of the necessary learning on their own, or with groups of other students, allowing their teachers to be spread more thinly. Most teachers in higher education already knew that students relied on each other a great deal to decode and make sense of lectures, reading and the like. Many of the issues about this particular project are addressed and discussed in the following case study. The dean of the Arts & Education faculty, for whom I was regularly doing curriculum design, course material writing, lecturing and dissertation supervision, asked me to be the faculty coordinator for this two-year-long project.

***

Case study The Student Centred Learning (SCL) initiative: the need for some theorising

(Any names have been anonymised, and references to the institution concerned removed.)

Most colleagues are now familiar with the aims and objectives of SCL. Certainly many people will have become aware of these current issues around learning, teaching and curriculum change from earlier, smaller projects. Most colleagues will now have seen and read the letter from the two deputy vice-chancellors (5.5.99), seen the blue SCL week leaflet, and attended a meeting where these ideas and issues have been aired.

Through talking to people I have noticed that, on the one hand many colleagues are 'new initiative weary' and yet very aware of the consequences of changes in university life that have a bearing on us all. The way all practice is clearly at the centre of most people's thoughts, and whatever reservations we may have about SCL as an initiative, it is undeniable that this myriad of issues is on our individual and collective agendas.

One of the objectives identified in the deputy VCs (vice chancellors) letter was to

> facilitate and encourage institution-wide debate on the implications of student-centred learning for subject, discipline or support area, and in particular to consider its potential impact on the curriculum, the structure and delivery of academic programmes, and research.

This 'debate' is already under way; that much is clear. But what is not so explicit is the role of theorising within, and for, this debate. We are after all working within, contributing to, a network of social relations, an organisation that values theory and theorists. Hopefully we also value theorising, which as a social theorist, I have always considered to be a/the key role. As reflective practitioners we are only too well aware of the diverse nature of our roles and the nature of interfaces with others. Most colleagues I talk to do regard themselves as research-minded practitioners, i.e. where the value of theorising is at the heart of their day-to-day practice, where research is like an intellectual journey, where we do constantly consider and evaluate our motives, aims, objectives, success at achieving what we claim and so on. We are also aware of the danger of 'packing our bags for these journeys, but never actually making them!' Part of the difficulty here is the disparity between the theories related to our practice as a . . . whatever, that we espouse, and our actual theory in practice.

Those of us engaged in teaching also have a further prevailing problem in that we have to juggle with a dual practice. We are practitioners in our academic disciplines, in my case sociology, but we are also practitioners of education. Even at the most basic level these practices, disciplines, academic and education, are not one and the same. For example the current body of knowledge paradigms are invariably separate, and very likely to be underpinned by fundamentally different theories of human action, and of research methodologies. To acknowledge these differences is the beginning of an intellectual, and essentially research-focused journey. We all make these journeys, deliberately or not, voluntarily or not, with an explicit recognition of these processes, or not!

Over many years I have noticed that 'academic' colleagues are proud to demonstrate their up-to-date-ness in their own discipline, and so they should. However, I have also been regularly struck by the out-of-touch-ness (not just not up to date) of colleagues when education as a practice is discussed. It has been the case where very able theorists have boasted of never ever having read a book on education theory!

Now, I am certainly not saying that it is impossible to be a competent teacher, or to understand learning, or be aware of the need for curriculum innovation, without a sound knowledge of education theory, but I am saying that we need to address these issues within SCL. To deny access to these intellectual concerns 'at the front door' means that they will only return unannounced elsewhere.

The SCL initiative quite rightly focuses attention upon the student experience, and the published aims and objectives make that explicit. My word of caution is that those of us who are engaged in the practice of education need to get 'our own house' in order along the way.

Not the least of the issues raised by SCL is that concerning the learning organisation. Now, in the context of H.E. this may seem tautological! However, in my experience I am not so sure. To put it briefly (if only!), a learning organisation is 'where people continually expand their capacity to create results they truly desire, where new and expansive patterns of thinking are nurtured, where collective aspiration is set free, and where people are continually learning how to learn together' (Senge 1990).

So if we are to embrace SCL, we also have to consider this. This may encourage some people to suggest that this is another sound reason for doing nothing, but what might be the implications of taking such action? It is clear that complicated issues to do with our individual and collective identity are bound up in these concerns. Our own sense of self and self-esteem are central to our practice(s). Our sets of values are under threat; our workaday principles are under siege and so on. And these are just the positive factors! To avoid being accused of sophistry, I would add that these issues do raise existential questions in that we know we should act, yet we feel that our actions are circumscribed by the contexts of our everyday lives, and to not act, or not act in harmony with our values would be to act in bad faith.

If we believe that our practice contributes to a learning organisation for our students, via the organisation of learning, do we also feel that this is true for ourselves? Are we as centred as we hope our students are or will be?

We need to explicitly use theory and theorising to unpack SCL. Perhaps a good place to start would be to ask what we mean by:

Student
Centred
And, learning?

At this point I would just add a few thoughts, very much my own 'work in progress'.

What sort of student are we talking about? The autonomous individual (or even the individual seeking to be autonomous) à la reflexivity or flexible accumulation, i.e. the student in the singular?

Or, the student in the plural: part of a community, where amongst other things authority is derived from mutual inter-recognition and reciprocity?

We are all free action-taking/choice-making persons, and social beings in the sense of being part of the collective. We are not for oneself as much as ourselves; our life development is not just endlessly self-referential but inter-related with others. We see ourselves through others eyes, and constantly seek to 'remake' ourselves through these interfaces and interactions. For me this is an aspect of andragogy, the nature of adult learning and teaching related to adult needs. This raises questions like, is an adult education (in respect of learners and teachers) to be liberation or domestication? While this may be a false dichotomy, i.e. nothing is ever that cut and dried, the issues are real enough.

Given these basic questions, what do our educational, learning and teaching processes, encourage now (e.g. via assessment)? What should we be encouraging? As practitioners we claim to deliver certain aspects of a curriculum (even broader educational goals as well), but what do we know of our effectiveness? If we (over)-rely on the measurement of our efficacy by conventional techniques do we know whether our practice is effective, even efficient, in our own terms or that of others?

What do we understand by centred? Most contemporary wisdom suggests that the dominant human characteristic is a de-centredness, a demonstrable lack of self-actualisation, individually and/or collectively. This phenomena is variously called alienation, anomie, the saturated self – all suggesting that we are not fulfilling our potential, because we are trying to lead a life within an arrangement of networks – economic, social, cultural, moral, psychic etc. –that actually prevents us from being fully human, really ourselves. And there is much more!

A good deal of the above comes back to a discussion of role and role inter-relations. What do we inherit when we take on a role like student, or teacher, or learning supporter, or . . . ? What is expected by us and by others? What is the nature of the relationships associated with these roles, e.g. power, trust needs? What can we hope to achieve in this role and how does this relate to our identity, our sense of self and others, and others' sense of us? Always bearing in mind of course that we are all multiple selves. Do we critically reflect upon these relationships within, and with others as a constantly changing process, often set against a desire to achieve some degree of certainty, of continuity and security in an uncertain, risky world?

Besides anything else, this can remind us that we are the subject of such reflection and not the objects that just get done to. How self-consciously reflective and analytical we are will perhaps greatly influence the nature of our choices when faced with the bold reality of our freedom to take action; and we have to do just that!

What then is the basis/motivations for our action/choice? Is it, are they, consistent, non-contradictory, in good faith, even virtuous? How would we know? Do we have a methodology to test this out? Is it explicit,'on the table' for all to see?

Every so often we all see, and recognise for what it is, the intersection of our own 'personal troubles' with the sets of social relations (including institutional ones) that are contextual to our lives, and, constantly changing. Do we make key decisions about our significant transition at such points of intersection and revelation? Is this transformational in that we make qualitative leaps forward at such times to become the updated versions of ourselves, the people we want to be?

There is now an enormous literature on learning. Inevitably in an era of 'publish or be damned meets marketing meets lack of quality control', a great deal of the literature is recycled. However, it is there, and a good deal of the key ideas have already been aired in the SCL initiative papers to date. It seems to me that one of our collective tasks is to clarify, and make some sense of, this extensive literature, demonstrating as it does the considerable twists and turns of theorising. A lot of these ideas are, as I have often said, more likely to clear the room, rather than clear the mind, but we do need to address these ideas.

One of the tricky questions is, do we keep all this theorising 'close to our chests', as the guardians of the knowledge, or do we engage with 'our' students in making these ideas an aspect of the transparent processes of education? Will they be grateful for our candour? Will they be empowered by becoming co-owners? And so on. Needless to say these questions are part and parcel of the learning theory debate.

In a more 'practical' way how can we make this literature, this knowledge, these ideological debates and so on, accessible for everyone? I have already been discussing with colleagues the establishment of an 'archive' that would keep the materials up to date by using it as 'work in progress'. There are other models and other means we need to discuss and develop.

We need to consider how to take the thinking about the SCL process forward. There is the SCL week to come in November, and the subject 'away-days', plus any other all inclusive school, departmental or faculty events. These various gatherings could be used to open up and widen dialogue.

In this very brief paper I have attempted to set out a case for the value of theorising at the heart of the SCL initiative. After all, this is a university!

As always I look forward to responses from colleagues.

John Astley 8/99

Senge, Peter, *The Fifth Discipline: The Art and Practice of the Learning Organization* 1990

***

In the nearly twenty years since I wrote the piece above most institutions of higher educated have adjusted to the routine of more students and less resources. Since fee-paying students have been turned into customers, their expectations of face-to-face contact with their tutors have also shifted, and there have been several examples of major complaints by bodies of students about the lack of tutor-student contact of all kinds.

In this chapter I have addressed a wide range of issues associated with designing and delivering curricula, and specifically so for professional practice courses. The role of social science in courses has certainly changed, with for example much less employment of social science specialists involved in all aspects of design and delivery. Social science concepts and bodies of specialist knowledge are however ubiquitous, still seen as necessary and valuable within the contemporary curriculum. However, the boundary around subject and module areas has become blurred. There is now a greater blending of social science knowledge with practice knowledge derived directly, but reflected upon, from practice professionals themselves. This is possibly due to their own educational experience where they became aware of the contribution of social science in an understanding of professions and practice, and became more confident in interpreting this knowledge in a more integrated approach to exactly what is in the curriculum relevant to a particular course, and who should deliver that.

I have also said a good deal about adult education and the shifting debates around an increasing focus on learning. Most research and theoretical work 'on' adult education and the adult education curriculum has in fact focused on learning, and on the nature of the student, the learner.

Re-thinking adult students, as, well, adults and not children, was a crucial step in thinking about learning. Nineteenth-century ideas about the autodidact have morphed into the concept of self-directed learners, which has in turn fuelled a movement for adult education in particular to be a transformative process for both the individual and society at large. If I was to cite one important example of how these three ideas came together and prospered, it would be the creation and development of the Open University in the late 1960s and early 1970s.

I have also said something about the importance of the 'social model' for health and social welfare practice, which not so surprisingly has had a profound effect on our understanding of the adult learner:

> [T]he *adult learner* is seen wholistically [*sic*] . . . the *learning process* is much more than the systematic acquisition of and storage of information. It is also making sense of our lives, transforming not just what we learn but the way we learn, and it is absorbing, imagining, intuiting, and learning informally with others. . . . [T]he *context* in which learning occurs has taken on greater importance.
>
> (Merriam 2001 p. 96)

A number of these issues about the adult sensibility and educational experience in general, and within professional education in particular, are explored and developed in the next chapter on roles. I will be discussing the generality of this key concept, but also considering how the roles of those designing and delivering courses for professional practice have also changed in significant ways. Many of these changes have come about because of the many and challenging ways in which social science has developed our understanding of the creation of roles in society, of role taking and role performance.

Chapter 4

On roles

Having spent time looking at the 'what' and the 'why' issues of the curriculum, it is now necessary to focus more specifically on the key issue of roles, and certainly so of professional roles. My aim here is to consider the taking of, and the performance of roles with a focus on the inter-relationship between the private and public aspects of roles and role taking.

In this taking on and performing of roles we should also acknowledge that all roles come with a set of rules and these inevitably create one aspect of the context affecting, and effecting, what a practitioner actually thinks and does. In general terms the 'rules' attached to each and every role in our lives are the product of a broad cross-section of cultural influences. Our current society is a liberal-democratic one, i.e. a 'liberal', even neo-liberal in contemporary parlance, free market, globalised capitalism, uncomfortably coupled with a representative parliamentary/legislature/government democracy of some historical convention – but also for the time being 'wrapped around' by membership of the EU. There are also some significant differences amongst the nations that comprise the UK. Scotland for example has significant differences in its legal and educational systems.

It is not my intention to enter into a lengthy discussion on the economic and political form of contemporary British society, but I do need to emphasise that for some time now the cultural norm, even mores, of this so-called liberal-democratic system is very well established. I did comment on these contextual issues in my introduction, reflecting upon the nature of civil society and so on. A good number of the role-related rules that have an influence on the day-to-day life of the professional practitioner do stem from laws about this and that. These laws, acts of parliament, have been enacted by the state and carry the authority of 'our' representatives. There are of course many other cultural groups that create and enforce rules in the everyday life of the professional, for example the professional associations, or societies which organise, support, do research for and represent professionals. There are also various statutory bodies created by legislation or other devices that oversee standards, inspect, examine and so on. There are also localised culture groups where local custom and practice is developed and maintained often via local protocols. All of these cultural groups have power in one form or another that has a history to it. They will also have some authority, and from a sociological perspective, authority = power + legitimation. As part of the acculturation process new members of a profession are encouraged to sign up to the cultural expectation of the 'group', and usually exercise very little agency in this process. The consequences of what they have taken on will eventually manifest, especially so when the practitioner turns the role holding into performance.

On the issue of taking on roles I usually asked my social work students why they wanted to help people. They often looked surprised that anyone would ask them that question, and probably confirmed in them the thought that sociologists are strange people.

One sociologist of the immediate post-war period who had a considerable influence on these issues was the American C. Wright Mills.

> By choosing social role as a major concept we are able to reconstruct the inner experience of the person as well as the institutions which make up an historical social structure.
>
> (Gerth and Mills 1954)

Mills' focus on the conceptual importance of role has resonated ever since, for example in the work of a fellow American, Richard Sennett cited earlier. Mills asserts a sociological truism, namely that we take on roles, with all their attendant rules, and perform these roles with inevitable psychological consequences for us as practitioners, shaping and challenging our character.

As argued above, the twentieth century was consistently referred to as 'the century of the self', where the myth of complete and fulfilling independence for the individual was promoted. In western capitalist societies this paradigm of the nature of the autonomous self, became one of the taken for granted characteristics of everyday life in the 'open society'. Most of the libertarian and freedom zeitgeist of the 1960s sought after, and revered and celebrated, this 'open-ness' in many diverse ways. This became in turn one key aspect of the cultural norm of contemporary society. The growth of possessive individualism has been a consequence.

I have already referred to the self and society tensions in everyday life that is a key issue for sociologists, and most socio-psychologists, in their understanding of people. I want to say a little more about the concept of self:

> The self is constituted and refashioned through reference to a person's own understandings, opinions, stocks of knowledge, cognition and emotion. The self cannot be articulated independently of such practical knowledge or consciousness, since self-interpretation enters into the fabrications of identity and the self in a chronic way. If it did not, the self could not survive nor adapt to changes in the social world.
>
> (Elliott 2001 p. 5)

Two immediate issues to take from Anthony Elliott are, firstly this point about interpretation. Sociologists like myself (and Giddens 1991 for example) have often pointed to the problem of the double hermeneutic in our understanding and analysis of social action. (Hermeneutic means the study of the principles and methods of interpretation.) What we do as sociologists is to interpret the actions of people, but they are already interpreting their own actions because as Elliott suggests we are all in the business of self-creation. So as sociologists we are interpreting the interpretation. The technical terms, used by sociologists and anthropologists, are that *emic* is the insider's view of 'a' culture, while *etic* is the outsider's view of the same phenomenon. But this situation is fluid as we are all also constantly re-inventing ourselves, telling stories to ourselves which may call on our unconscious desire, or dreams and fantasies, our utopian visions. The current 'Me Too' social media fad is in general terms one derivative of all this. We all have a locked 'filing cupboard' which we look at occasionally and decide how much of this stuff, if any, we can disclose? This is very much part of the private domain of our lives. Then have to present this self in the public domain, and this is particularly true when we are taking on and performing roles. We should add to this already present weight of speculation about the self the apparent growth in what is called identity politics. We now have an extensive discussion in society about who wants to be called what, and who can call them that. This is another key reason why for the social sciences in general, and certainly for me as a sociologist, there is a preoccupation with the humanities in all its diversity including the arts, music, spirituality and so on. It is also crucial that we understand the significant context here of the *modern* world, the world of sciences, of the growth of a global industrial capitalism, of (allegedly) rational man. Sociologists, certainly this one, continue to engage in lengthy philosophical 'arm wrestling' with the outcomes of the Enlightenment on our thinking about the human condition, for good or for ill. We all want to be at the least semi-autonomous beings, with agency, in control of our own choices. Perhaps?

> Autobiographies are made of personal memories, the sum total of our life experiences, including the experiences of the plans we have made for the future, specific or vague. Autobiographical selves are autobiographies made conscious. They draw on the entire compass of our memorised history, recent as well as remote. The social experiences of which we are a part, or wish we were, are included in that history, and so are memories that describe the most refined among our emotional experiences, namely, those that might qualify as spiritual.
>
> (Damasio 2012 p. 210)

However, in the meantime in the 'real world' we all need to deal with the everyday paradoxes of being our genuine thought-through self in a society still stratified along rigid class, gender and ethnic lines. Being a role holder continues to place many demands on us that are antithetical to seeing freedom for all alongside freedom for each. Right across the texture of everyday life, from family relationships to the workplace, we invariably struggle to make sense of it all, the expectations and the disappointments, and often end up being frustrated.

There is an echo here again of John Gray, the political philosopher, when we argue that 'everything' good has a shadow to it, the downside, the not-yet-complete, work in progress.

> Usually, we can accept that what we desire does not always exist; that we can desire something without necessarily hoping for it and certainly without expecting it. But perhaps there might be areas in our lives where we can believe something to be the case because we desire it, and find ourselves unable to give up the idea whatever evidence we are presented with.
>
> (Craib 1994 p. 4)

In this last respect I would highlight the potential pleasure that giving service, helping, caring and mentoring can bestow on us, acting as an antidote to the many frustrations to be found as a professional role holder. Our values can be reinforced, and we can certainly have confirmed the belief that practice can be a virtuous circle. An issue I discussed with those very social work students mentioned above.

Mills was developing his work on roles to illuminate the key issue I touched on above, namely the constant but complex inter-relationship between the self and society. In his classic text *The Sociological Imagination* of 1959, Mills exhorted his readers to develop an imaginative grasp of the conceptual pairing of personal troubles and social issues. And 'imagination' is crucial here because any attempt at empathy with clients or colleagues requires an imaginative leap 'into their shoes'. A key aspect of this approach is in the balance of likelihood of life events such as unemployment, acute illness, an unsatisfactory schooling experience or divorce. Mills suggests that if he was the only person in a hundred who was unemployed, that certainly says something about him, his personal trouble, even his fault or lifestyle choice. However, if he is one of several people unemployed, this would then be seen as a social and public issue, and even likely to attract the attention of policy makers and even legislators. This framework for thinking about the inter-relationships of everyday life also alerts us to what has been all too common, the contrast of private affluence and public squalor – another aspect of the unfairness which has been increasingly noted by people living in contemporary UK society.

Here we can see both the utopian dimension of an 'imaginative' grasp with the constraining and compromising realities.

Mills' aim was to help create a way of thinking about personal troubles that transforms them into public issues, placing/forcing them on to a political agenda, and thereby emphasising the ideological character of social policy often lost at the micro level.

It could be argued that his critique, exposing the contradictions of the everyday opened up a discourse that challenged the hegemony of the cultural norms of his and our time. The compare-and-contrast notion of exposing the 'snakes and ladders' of opportunity and consequences in everyday life has been widely used in professional education to emphasise the personal, and yet collective, nature of what is often rather abstract discussions about 'social problems' and the individual.

This conceptual focus encourages us to consider the plight, reaction and agency of that one person, who is clearly a social being, but also an individual who deserves, even has a right to, our consideration, compassion, empathy and support. A good deal of the ethical arguments around practice in the welfare domain emphasise this point. Indeed the 'golden rule' of ethics is often evoked, namely that we should do unto others what we would want for ourselves. In their attempt to grasp the everyday realities of an existence with endless contradictions, false starts, missed opportunity and so on for most people, social scientists seek to identify and explain the range of factors that can and do have a bearing on standard of living and quality of life. Mills also encourages us to consider the creation, development and threats to character in a person. Who is likely to cope or not given a certain set of circumstances?

Since Margaret Thatcher declared that 'there is no such thing as society' (in a 1987 interview with *Women's Own* on who has responsibility for the welfare of children and young people) we have continued to struggle to rebuild any viable sense of the reciprocity required to maintain and develop civic responsibility. The power grab of the centralised state has stripped local government of the resources, authority and political will to challenge increasing inequality in all respects.

> If we are to pursue the sociological agenda in terms of a social critique . . . if we are to compare and contrast the culturally effective conceptions of a successful life with the social conditions under which they are pursued, then the direction that a contemporary social critique must take is fairly evident. The aspiration to autonomy so characteristic of modernity, the ideal of a way of life independent of material and economic constraints, is being continually and ever more intensively frustrated – at a collective level of political organisation of society (a) as well as at the level of individual life conduct (b). At the same time, this leads to experiences of alienation (c) and moreover, is linked to dysfunctional effects (d) which endanger the reproduction of the system independently of the cultural ideals that serve as its base.
>
> (Dorre 2015 p. 89)

This is a further reminder of the consistent paradox of being 'freedom's prisoner', and given the vicissitudes of everyday life for clients, client groups and practitioners. Is there an appropriate response from education providers?

In many instances curriculum designers have actually changed the nature of knowledges on offer by broadening the methodologies used to collect, analyse and discuss experience. For example the use of client (auto)biographical evidence has enabled students/practitioners to consider first-hand accounts of everyday life that would often only be filtered through 'academic' texts. I often made a point of drawing from my students their personal experiences of being social welfare clients to compare and contrast with the knowledges they were also considering via the conventional academic route, usually provided by me either directly or via the reading list.

I remember well just such an occasion when discussing the vagaries of clients trying to cope with officialdom in addition to dealing with pressing personal and/or family issues with a group of social work students, when one student in his forties discussed his 'textbook' experience of precisely these phenomena as a recovering drug addict. His fellow students listened very carefully to what he had to say.

There is here an example of how some methodological approaches, like ethnomethodology (the methods that 'the folk' use to make sense of their lives) and oral history, meet with critical theory to challenge practice orthodoxies. This is not to suggest that by simply experiencing some social phenomenon ourselves we will fully understand the complexities involved. This is why theory, especially so in a comparative evidence-based dimension, is crucial to developing a critique of any social phenomenon and the interpersonal relations contingent to that. However, we as educationalists supposedly designing and delivering courses related to the 'real world' in which we live, should be prepared to know about, and even work with the many community and issues-based pressure groups that work in our locality. We may, as I certainly tried to do, be actually working with such groups in some capacity.

In the last decade the concept of the 'precariat' (a class of people with particularly precarious lives; more snakes and ladders) has become part of the discourse on vulnerability and needs. The increase in the number of people with precarious lives has put increasing strains on welfare practitioners, especially so when faced with an ideologically led set of 'austerity' cuts to resources both human and material. And to follow Mills it could be argued that to prepare professional practitioners to enter into social interactions of many forms without a well-developed understanding of themselves as adults engaged with other adults would be inappropriate to say the least.

Erving Goffman was one the sociologists in the generation after Mills who developed ideas on roles. He argued that as professional practitioners we need to pay close attention to the way in which our role affected our engagement with clients, and the effects that such encounters actually had on those clients. For example, clients might have preconceptions and past experience about the nature of people like us. As 'people workers' practitioners are constantly engaged in face-to-face work, the place and time of 'interface' between people, who may be in this situation because of shared goals. Recent bureaucratic initiatives would wish however to place practitioners on a telephone looking at a computer screen. This is a further dimension to the increasing power of new IT-led technologies in all of our lives. (See Byung-Chul Han on '*Psychopolitics*' 2017, amongst others) This de-humanising approach to service and care delivery is no mere contrivance to cut costs; it is a deliberate attempt to remove practitioners from the direct personal contact with clients, and their families, which is likely to stimulate empathy. Out of sight is out of mind. Another outcome of such interaction could be an aim to enable clients to be change agents in their own lives, and create in professionals a desire to radicalise practice and the systems that contextualise practice.

Goffman's work was diverse and covered many aspects of the human condition and the nature of social interaction, including his fascination with the highly charged symbolic world of language. He did also, especially in his earlier work, write about the roles of practitioners, and clients in relation to practitioners, for example in his collection of essays titled 'Asylums: essays on the social situation of mental patients and other inmates' (1968) and 'Stigma: notes on the management of spoiled identity' (1968). I would recommend these to any professional practitioner, new or slightly shop-soiled.

In consideration of the 'making sense of ourselves' focus in this book, let me quote Goffman on this:

> [B]elief in the part one is playing. When an individual plays a part he implicitly requests his observers to take seriously the impression that is fostered before them. They are asked to believe that the character they see actually possesses the attributes he appears to possess, that the task he performs will have the consequences that are implicitly claimed for it, and that, in general, matters are what they appear to be. In line with this, there is the popular view that the individual offers his performance and puts on his show 'for the benefit of other people'. It will be convenient to being a consideration of performance by turning the question around and looking at the individuals own belief in the impression of reality that he attempts to engender in those among whom he finds himself.
>
> (Goffman 1959 p. 15)

This sounds straightforward enough, but as teachers/lecturers/presenters/talkers engaged in presentations all manner of symbolic issues can get in the way of understanding, and often misunderstanding. I recall doing a talk on William Morris to a group of art college design students. I wanted to place the focus of my talk on the enduring characteristics of Morris' design, particularly his 'naturalistic' approach. As I had often done before I took with me to use as visual aids a quantity of fabrics, wallpaper and other visual materials, and had in fact put all this into a large Debenhams store bag. After the usual introductions I gave my illustrated talk and did a Q & A session; all went very well. The lecturer who had invited me called a few days later to say how much they had all enjoyed my talk; one of her students commented that it was amazing how much the man from Debenhams knew about William Morris. I was pleased with that.

One consequence of these situations is to take us into the realms of values, and the inter-relation between roles and values. In all my work with professional practitioners I have sought to open up a discussion on why values do matter for everyday practice. If several decades of practicing as a sociologist has taught me anything, it is that of all the concepts we have in our 'intellectual toolbox' values is probably the most important. A focus on personal and collective values can help us to develop an appreciation of the complexity of professional life, including coming to comprehend the diverse aspects of top-down micro-management-motivated social control. These range over surveillance techniques and computer-mediated communications in an increasingly virtual world. Our values can act as an alarm call to help shape our actions and responses to others drawn into this complex and obfuscating set of social interactions. This sounds positive, but the downside here is frustration, disappointment and disillusionment. Increasing numbers of professionals of all kinds feel that their ability to act in line with their values is undermined with a consequent loss of integrity, and low morale is one common consequence.

Recently I was invited to facilitate and deliver a two-day staff development event for a major voluntary organisation in the south-west of England. I was asked to focus on values, and how an understanding of the values held by the staff could lead to a wider and deeper appreciation of their roles and relationships with colleagues and their clients.

I have reproduced below the pre-event 'flyer' that was sent to all participants, and which I used and amplified throughout the workshops.

***

Why values matter for our everyday practice

In my brief talk and workshops I wish to focus our attention on four key issues for us all as professional practitioners. These are VALUES, ROLES, CONTEXTS and REFLECTION in action.

I will argue that we need to understand how we acquire the VALUES we have, and values which undoubtedly underpin our role taking and performance. Which values are most significant for us: empathy, altruism, compassion? What are the opportunities that holding the values that we do give us, and to our clients and colleagues? What are the threats to our values, and our desire to act upon them?

Secondly we need to explore the CONTEXTS to our roles and role performance. Why are we doing these jobs? Why are they necessary in contemporary British society? By taking on the roles that we do are we constantly 'swimming against the stream'? For example, are contemporary social welfare policies part of a solution to needs, or part of the problem? In these circumstances can we be change agents, for ourselves and for others?

Thirdly we should consider the nature of the ROLES that we have taken on. All roles come with a set of rules, for example organisational and legal ones. But our own values, attitudes and prejudices also act as a set of moral rules, or codes, to guide our actions. Are there conflicts here, and how do these issues impact upon our clients and colleagues? After all it is said these roles exist to provide a service that is needed. That is why we are doing them, isn't it? Can we hope that what we do, and how we do it, gives us a sense of purpose and worth, and crucially, lead to a virtuous practice?

Lastly we can discuss our place within what is usually called REFLECTIVE PRACTICE: briefly put, the ways in which we assess the effectiveness and efficiency of our everyday practice. How does this 'reflection' happen: are we reflectors in action, regularly testing out our claims to be doing a good job? What techniques do we use to ensure that practice is under review? Ours, others', clients'? There are both governance issues here, and the wish to develop our reflexive selves into research-minded and critical practitioners.

Once we begin to explore the motives behind, and contexts to, our everyday actions we enter a moral, ethical, emotional and institutional 'minefield'. Is our 'intellectual toolbox' in good enough order to deal with this?

In this workshop session we can discuss these issues, and much more!

John Astley

***

And we did, and with the usual mixed reception from the participants engaging with these issues.

I would argue that when focusing on the importance of exposing and discussing our values, the context of acculturation looms large. We all hold a set of core values, but how did we acquire them? What agents of socialisation and social control sought to domesticate or liberate us? When a person sets out on the journey that is professional education, what do they bring to the table, including their values, and are they prepared to be challenged every step of the way on that journey? The importance of reflecting upon the roles we take on, and perform, is that there is always both a personal and a public dimension to roles. A key part of any professional education should include a full and frank discussion of why a certain set of values has found its way into the practical reality of the curriculum. It is commonplace for students to have their understanding of the importance of values and roles, and all their contexts, challenged within the educational setting (the practicum) only to find that the workplaces that they then access during and after their course are less comfortable with such critique. Students are, and should be, encouraged to see reflection in action in their actual day-to-day practice as essential to developing an understanding of their role. Students do need to be warned about this issue, and helped to prepare themselves for the possible struggle ahead. The educational process needs to give students the confidence to challenge practice orthodoxies, an entitlement to be contrarian in whatever settings they find themselves.

> Underpinning the effective and purposeful teaching and learning of young children is the engagement by early years practitioners in critical reflection, and directly related to this is the active and meaningful participation of children in what they do. . . . What becomes quickly apparent when listening to effective early years practitioners and primary teachers is the nature of the conversations that they have talking about the children they work with . . . They take great care making inferences based on observations and they continually look for new meanings in what they do.
>
> (Gray and MacBlain 2012 p. 141/2)

In response to the interesting point that Gray and MacBlain make I would like to include here two pieces written by recent ex-students of mine, who take up these issues well.

When I first began to study at degree level I was struck by the wealth of information and research that was available that both challenged or supported my own pre-existing beliefs and prejudices.

The process of study at degree level is designed to make you more aware of current thinking and developments. While this acted as a very useful catalyst for my academic thinking and writing it also presented significant problems and challenges in my working life in school. These mostly centred on the way I now viewed existing policy, working practices and the implementation of support in my setting. Having been exposed to new ideas and theories it was hard to see them being either ignored or cast aside in my work setting.

While understanding that there were factors such as cost of implementation and staffing, I was still often left feeling frustrated and angry that opportunities to improve outcomes for individuals were being missed. There seemed to me that there was reluctance on the part of my colleagues to hear my voice and to take my ideas and suggestions seriously. Initially I viewed this apparent rejection of new ideas and approaches personally. However, as my studies progressed I was able to acknowledge and respect those in my own organisation, and more widely, responsible for making, what to me often seemed, the impossible decisions involved in selecting who gets the help. They, like me, were only trying to make the best of the system that they had to work within. They, like me were informed by their own pre-existing beliefs and prejudices.

Once I was able recognise this it led me to the realisation that my best chance of influencing these decisions was by deepening my own knowledge of what was available and improving my professional practice. In this way I have increasingly been able to make relevant and timely contributions to the discussions in my setting, while still being mindful of the pressures that are in play for all of those involved in the funding, resourcing and delivery of interventions. The bottom line is that we are all striving to make the best use of what we have and by being better informed we stand a greater chance of making the right choices.

For the most part my colleagues were very supportive of my decision to undertake the degree. A few were surprised and wondered what I hoped to gain from it. However it quickly became apparent that not all of my colleagues would feel able to offer me practical help or support. For this group the idea of being interviewed or allowing their lessons to be observed proved to be more than they felt able to offer. Others were prepared to help provided they could see a direct benefit for themselves. Most though were very generous with their time and allowed me free access to any information I required.

I had one slightly difficult situation in my second year when I allowed my Head Teacher to influence my choice of research topic. This resulted in the production of one of my less successful assignments. The problem being, I felt on reflection, that I did not really 'own' the project.

In regard to my involvement in the provision of assessment and support for pupils with Special Educational Needs and Disabilities my colleagues are now aware that my knowledge is in most case more up-to-date than their own. This has resulted in my opinion being asked for more often. While this is gratifying it does not necessarily follow that what say is acted upon. There are also times when my contributions are viewed as adding to the problems rather than being seen as offering a solution. This, in turn brings with it feelings of frustration that my voice is not being heard.

(Mann 2018)

And,

Having worked in Early Years Education for over ten years, I have a great passion for the way in which children learn and how we as adults can support them to achieve to the best of their ability. My current role as play leader is a role that will strongly support me throughout this degree as I feel it will enable me to use the necessary skills I have developed over the years. I felt that I am at the right place in my personal life to be able to move my career forward.

For the first few weeks of starting this degree I wasn't sure what to expect. I did feel myself wanting to run before I could walk, always wondering when we would be given the first assignment, what would the content be? And how much would I have to write? The questions just kept popping into my head. I found myself feeling quite anxious and I didn't really understand what would be expected of me and what level of work I was going to have to produce. All these thoughts were spiralling and I did begin to question whether or not I had made the right decision. I realised I was going to have to take a look at the way in which I managed my time and to stop myself from trying to do everything all at once. I know this is one of my biggest flaws as I find it extremely hard to ask for help. At this point I took strength from my family, especially my husband, who helped me to organise my time. I also talked to other students who helped me to clarify the thoughts I was having.

The impact I am hoping the degree will have on me is the immense pride in myself completing a 'degree', something I have always wanted to achieve. It has become quite surreal to finally be doing this course and learning to balance life has been sometimes very overwhelming.

I am sure this will boost my self-confidence and by completing the degree it will enhance my professional development.

As we progressed through the degree, primarily level four and five, I found the modules relevant to the line of practice I was working within. The research undertaken allowed me to understand the theory behind the practice. This was really illuminating for me as I felt I could make more sense of the way children learn and develop.

Now that the degree is complete I feel that there was not enough opportunity to explore the possibilities of what next. I am a graduate with a second-class degree but within my chosen work place I feel I would have to complete a post graduate qualification for my degree to mean anything. This left me feeling quite deflated and when asking for career direction found the training providers were mot very helpful. However, I intend to upskill my literacy and numeracy and hope to complete a PGCE and finally graduate as an Early Years teacher within the next three years.

(Burns 2018)

One aspect of this issue is the creative potential that people will bring to membership of the professions, and to practice. Both the discussion about the roles we take on, and on the valuing of the attributes that people bring to initial professional education, are important. The rules associated with roles are open to challenge by creative minds and the life experience that people bring to their situation. People in all roles in any organisation need to ask themselves just what value is placed upon creativity, innovation and democracy. Is there a 'climate' for experiment, the encouragement of enterprising thinking? Is this a 'learning organisation'? Is the link between the experiential and the experimental sought and valued? And if not, why not? What are the barriers to potential cultural changes?

In 2000 the educational researchers Andy Hannan and Harold Silver based at Plymouth University published a report on Innovations in Teaching and Learning in Higher Education. This was a wide ranging project looking both 'old' and 'new' institutions. They looked at the prospect for and support of innovators.

> One strong message that, whatever the policies and good intentions, they can be frustrated by the structures, attitudes and the weight of history. However, powerful the changes imposed from outside or chosen from within, histories count.

They also emphasised that innovation went on in many varied circumstances, but were much more likely to happen and be successful where the institutional 'climate' was of a 'learning organisation' kind.

A focus on the centrality of values can be found in sociological writing over a long period, for example the work of Vance Packard with *The Waste Makers* (1960), with the book's dedication:

> To my Mother and Father who never confused the possession of goods with the good life.

to David Riesman with *Abundance for What?* on mid-1950s USA:

> We are coasting psychologically on the remaining gaps and deficiencies in the ever rising "standard package" of consumer goods but, beyond that, we have very few goals, either individually or socially.
>
> (Riesman 1964 p. 299)

to Ivan Illich in 1971 from his book *Deschooling Society* on the over-institutionalisation of compulsory education:

> Man must choose whether to be rich in things, or in the freedom to use them.

to Richard Sennett in his 1998 book, *The Corrosion of Character*, where he considers the legacy of a consumer-led capitalism:

> Character particularly focuses upon the long-term aspect of our emotional experience. Character is expressed by loyalty and mutual commitment, or through the pursuit of long-term goals, or by the practice of delayed gratification for the sake of a future end. Out of the confusion of sentiments in which we all dwell at any particular moment, we seek to save and sustain some; these sentiments will serve our characters. Character concerns personal traits which we value in ourselves and for which we seek to be valued by others.
>
> (Sennett 1998 p. 10)

There are echoes here of Tawney and his discussion of *The Acquisitive Society*, his 1920 book offering a critique of the selfish individualism of modern capitalist industrial society.

One dimension of practice that social scientists would want to emphasise is the inter-relationship between accessing and prioritising specialist bodies of knowledge, reflecting upon everyday encounters with clients, and colleagues, and addressing the reflexivity associated with being a research-minded practitioner. There is an assumption here that as a consequence of these social interactions a practitioner is encouraged to examine any given body of knowledge pertinent to their practice and be analytical. Pick it apart, and then reconstruct it, on the basis of reflection-in-action. This is fortunately how bodies of knowledge, and certainty so 'theory' are, as discussed earlier, turned into a manageable, relevant and fit-for-purpose resource.

Talking about theory should open up debate, not shut it off. Any social scientific theory is work in progress, and therefore theorising is where the focus should be. This is particularly important for students of social welfare–oriented professional practice for all the reasons suggested above, confronted as they are with shifting, yet persistently influential ideologies of the human condition with all its vagaries, for social scientists theories are interim propositions to be tested out in a real world. Does this theory help? Is this or that theory essential to my 'intellectual toolbox'? Whose interests does this theory seem to serve? A firm grasp of the essentials of theory as bodies of knowledge both at a schematic level of context, and a specific focus, a 'spotlight' on the particular is necessary, but it is the potential application of such, the use-ability, that matters.

This is where the essential tenets of critical theory are valuable because of the fundamental wish to ask difficult questions, to challenge the status quo, conventional wisdoms, to be contrarian and enter the fray bearing alternative ideas to bear on the discussion at hand. This situation also touches on the social scientific concern with society as communication where constantly engaging in a symbol-laden world can make us aware of the 'vocabularies of motives' of all those involved. How do we use certain kinds of language to convey messages and meanings about situations, conditions, actions to be taken, or not, and so on? What are the conventional, dominant, forms of discourse? How are we talking to each other about this or that issue, this or that person? Clients, claimants, spongers, scroungers, deserving, undeserving and so on. In our conversations with all and varied stakeholders in any situation using the most appropriate language and manner, both verbal and non-verbal, to convey what can be abstract and complicated ideas is an absolute skill for all practitioners, and a key aspect of role performance. An important part in the education of professional practitioners is to develop the skill of making their communicative encounters meaningful for all. Being conscious of cultural diversities requires a grounded aesthetic, a feet-on-the-ground-ness that seeks to be effective and supportive at the same time. Practitioners can become reflexive conduits for the transmission of ideas and information.

From my own practice experience a good example would draw on the two years or so that I spent as a volunteer children's and young person's advocate. As an advocate I had to be the voice of that young person, often quite literally at a case conference/meeting where my role was to say the words that they could or would not say themselves in that adult, professional and often highly jargonised setting. I then had to 'translate' what had been said, and by whom, to enable that young person – usually the client, after all! – could make sense of it and make decisions, exercise agency. Professional practitioners often find themselves in a similar role with clients of all ages and dispositions, regularly acting as conduits, even as an interpreter and extemporiser, to take conversation, ideas and understandings 'over the bridge of meaning' to make sense of life.

Not every budding practitioner is a 'natural' in these respects, but these are learnt skills, and therefore educators need to see this as a priority from the outset. Social scientific understandings can inform and guide here.

This broader aspect to professional encounters is therefore crucially contextual to what form social interactions take. How often do we hear about, or experience ourselves, 'things getting off on the wrong foot'? All social interaction is about information exchange, for example about the substance of a message, or about the *messenger*. It is evident that the power that professionals have, an aspect of the care *and* control norm, should lead practitioners to concentrate on the expression and control of information given.

I once gave a talk to medical students on how the initial 'physical' examination that doctors make of patients makes use of the formers' senses – touch, sight, hearing and so on – followed by other forms of interaction as the triage process. Their clients/patients are the subjects of this sensuous experience, not merely physical and inert objects. Medics like other professionals are invariably seen as authoritative, even authoritarian (the white coat syndrome) and this can impact on the nature of the social interactions that take place between practitioner and client. The presentation of self is here (à la Goffman 1959), and emphasises the necessity for reflection-in-action.

So, given this we should all, teachers and learners alike, reflect upon the question of authority. If clients, or significant others, believe the use of power (and even the role holding of that power) by professionals is not a legitimate exercise of the power bestowed upon them as a consequence of their role (and the assumption of appropriate knowledge holding), then authority is undermined and usually challenged, or advice disregarded.

These are all aspects of the way our roles are fixed and yet fluid enough to make adjustments based on our reflection-in-action, our experience of the role.

Decades of inculcation into a professional culture group have certainly affected the development of practice paradigms, the conventional wisdom on how to think and behave. All culture groups create symbolic boundaries to safeguard and enhance the mystique of that group, and often for the purpose of advancing that group in the hoped-for elevation of occupational and personal status within the division of labour. Hubris as an aid to advancement meets the vanity of small differences. (See Anthony P. Cohen's *The Symbolic Construction of Community* 1985 for a wider discussion of this issue.)

Professional culture groups are 'knowable communities', and 'communities of meaning' i.e. they are material culture, but also symbolic. The new recruit, the novice, enters into a acculturation process, and it is not surprising that within most professions there is a 'house style' of how things are seen, discussed or not, and done. The so-called 'glass ceiling' situation in most professional groups is rarely an 'accident' but usually the consequence of maintaining the gender and ethnic status quo. Received ideas, as the basis for day-to-day actions, are very powerful in any organisation. This is an all too evident expression of how power actually works in culture groups, and of the historical sets of social relationships that comprise those groups. Sue and Neil Thompson touch on this issue in their book on reflective practice, where they emphasise an 'anti-learning' organisational culture that is far too often found that acts as a significant barrier to innovation in thought and deed (Thompson and Thompson 2008). The roles we take on as professionals will be heavily laden with all these status characteristics.

The educators of professional practitioners, whether also members of that particular cultural group or not, have a key role in exposing the history of, and potential for, self-serving reproduction of this quotidian. The very language that is spoken in everyday social interactions is an aspect of the 'vocabulary of motives' issue mentioned above. While engaged in the process of familiarising the new recruit with both context and role specificity, the educator will need to make apparent the potential for the onset of myopia. At every stage of the educational programme the educator has a duty to require the practitioner to 'step outside' of this socialisation experience and view their intended professional culture group as a stranger might. The practice of terrifying new recruits into submission has surely passed, hasn't it?

One of the intended, or unintended, consequences of the de-professionalisation of practitioners by the state apparatus over recent years has been to accentuate the nature and role of professional values in order to change them. Many of the more experienced professional practitioners throughout society have found themselves reformulated and managed into 'strangers' within their own profession, challenging their integrity amongst much else. The sense of irony in all this has not been missed!

This line of argument returns me to an issue raised in my introduction in suggesting that unlike ideologues social scientists live by reason. However, as I have already indicated, maintaining this ideal is very difficult in the everyday battles over what knowledge counts in decision making over social policy. We know that social scientists labour long into the night to create reason-based bodies of knowledge that hopefully lead to explanation and understanding. This is what knowledge is about, and that it should be for all, held and shared in the public domain. Epistemology is that branch of philosophy that seeks to investigate what we can know, but this is far from being purely an abstract set of speculations. The limit to what we can know is constantly circumscribed by persons who seek to create boundaries, to deliberately limit what 'the others' can know. This is an aspect of the processes of creating and exercising power, and the form that education takes in any one era is a good example of this phenomenon. What I discussed earlier in regard to the methodologies used for collecting evidence of everyday life is relevant here.

The knowledge created by many, or even most social scientists, is an outcome of the desire to provide up-to-date and accurate information which can assist discussion, and aid decision making, including those made by the clients of professional role holders, opening up a space where questions can be formulated and answers sought. From each according to their means, and to each according to their needs. These knowledges and their dissemination should be resources for hope, and promote the democratisation of decision making and role performance. This idea does of course underpin the 'service to society' ethos central to debates about the nature *and* role of the professions in society, which have been happening since the nineteenth century. Indeed, one of the justifications that the state apparatus offered in mitigation for taking over the running of people's lives was that public/civil servants behaved in an objective and non-partisan manner and therefore were above any petty ideological interests – a further expression of the 'rational man' of modern society.

Some people still believe that, and when listening to the neo-liberal/conservative advocates for small government (a code for the further commercialisation of the public sector) squaring up to the enthusiasts for more intervention (more officialdom in our lives), we should be careful what we wish for. Certainly the advocates of the latter approach to managing society should carefully consider the top-down, undemocratic and often patronising reality of much of what most people experienced after 1945–48.

There is a lively discussion currently taking place here on the nature and role of social enterprises (SE), and how this fits with both the statutory and the voluntary sectors' social welfare provision. The advocates of SE argue that there are more localised, focused and small-scale options available to meet needs in a community. Embedded providers, with a grounded aesthetic mentioned earlier, are likely to have greater empathy with their neighbours and locality. These approaches could help to avoid the large-scale bureaucratic 'one size fits all' approach to meeting needs from both public and commercial sectors, and is likely to be more cost-effective, and hopefully more directly accountable, a bonus for better governance.

I will be saying more about these developments in Chapter 5.

Not the least of current pressures on social science is successive government determination to marketise education. (Illich ignored!) One of the recent fads has been to make higher education more responsive to the needs of 'employers', although in practice it is often difficult to understand exactly what those needs are beyond a cheap and compliant labour force. There is a lot of talk about skills, but nobody appears to be doing anything about a serious policy and practice approach to skills-based vocational education. Further and adult education has been decimated, and the state and their commercial sector allies seem devoid of ideas and initiative.

This situation is even more complicated if those employers are in the charity and voluntary not-for-profit sector. What government has been suggesting is in fact pre- and in-service training that bears little relation to the semi-autonomous approach that higher education, and certainly a social science education, has offered. I have argued above that a social science component in the education of professional practitioners should enlighten participants to the complexity of society, and their role in it, and equip future practitioners with insights and skills to address issues and needs for all stakeholders. The instrumental attitudes that dominate further and higher education currently are antipathetic to speculation on the suitability of conventional bodies of knowledge. The neo-liberal ideology–led 'business model' has most certainly turned tertiary education into businesses first and foremost, to the detriment of educational values and the integrity of practitioners.

Those of us who have been involved for some time in the education of professional practitioners are aware of the great pressures upon new members of the various vocational culture groups. The advance of a state-sponsored technical rationality has led to a 'box-ticking' mentality, eliminating as much as possible from assessment and provision the inconvenience of real people with diverse lives, needs and expectations. Of course the education of professionals has been vocationally oriented. Social science teaching has, in my experience, been focused on the practical and necessary outcomes for people pursuing their education, and their subsequent lives as employed practitioners of various kinds. However, in recent years the almost sole emphasis on securing qualifications has changed matters. It is quite evident that for many students education gets in the way of their education.

As I have indicated above one key strategy that we social scientists can employ, and pass on in our struggle against the unreasonable ones, is emancipatory research.

> [E]mancipatory action research presents criteria for the evaluation of practice in relation to communication, decision making and the work of education. It provides a means by which teachers can organise themselves as communities of enquirers, organising their own enlightenment. This is a uniquely educational task – emancipatory action research is itself an educational process.
>
> (Carr and Kemmis 1986 p. 221)

As I have argued above, there is nothing particularly new in what Carr and Kemmis say, but what needs to be said is that 'we' are the subjects of our own research on action, and in our own action research. We also need to remind ourselves for whom all this expensive and time-consuming activity is supposed to be for. So beware of any tendency to solipsism in the quest to make sense of ourselves! How altruistic are we in our day-to-day practice as social scientists and educators? I still come across examples of research where those peoples being researched are the objects, and not the subjects, of the research process.

Hart and Bond expand on the nature of action research in their essay, 'Using action research', to be found in Gomm and Davies' excellent book, *Using Evidence in Health and Social Care* (2000).

Hart and Bond identify a number of features of action research:

1 It is educative;
2 Deals with individuals as members of social groups;
3 Focuses on problems in their social context;
4 Is informed by a cyclical framework;
5 In involves a change intervention in which there is a complex interplay between research and action;
6 Is concerned with involvement and improvement; and
7 Is collaborative and involves participants in the process as change agents.

They also elaborate the 'cyclical framework' listed at item 4 above, namely the process they call 'research into action', of moving from 'diagnosing' – identifying the problem – to 'action planning'; selecting interventions to 'action taking'; implementing change to 'evaluating'; reviewing the consequences to 'specifying learning'; exploring general findings, and so on, through the cycle again. At every point of this process we could be asking ourselves and other selves: Whose interests are being served by choosing one option over another? Whose voices are we hearing and listening to? Why select one intervention over another? Are 'we' thinking about both the intended and unintended consequences of any action being taken? So, who are drivers of the research, and what are the desired outcomes? Are they in fact formative or summative?

Hart and Bond suggest that there are at least four types of action research: organisational, professionalising, empowering and experimental. The experimental touches on the question posing/critical response to problems via a social scientific approach, the 'bread and butter' basis of science. The organisational type is largely related to an early response to evidence from the client base, and taking research incentives from that. The professionalising reflects the way in which professions seek to develop and improve on their existing bodies of knowledge, drawing directly on practice experience. Finally, the empowering is closely linked with community working, for example seeking to build resilience into a specific community and /or culture group in order to anticipate and address issues like health or housing (Hart and Bond 2000 p. 98).

Whatever approach may be taken, there are still key questions about the evaluation of any research programme. Social scientists in general hold the view that they could, and should, play an effective part in any evaluation, including setting up and monitoring such a process. (For an extensive discussion of these issues see Everitt and Hardiker 1996.)

There are still some social scientists that hold to the 'value-free' ideal of playing a neutral role in research programmes, including evaluation. This claim to a 'scientific' type objectivity has been considered and found wanting by most contemporary social scientists. There is in fact a long history of social scientists who have chosen to be embedded within a community for the purposes of research, whether as a 'freelance' or part of a sponsored scheme, and argue that transparency of motives and sharing of process, progress and outcomes is ethically essential. It could be argued that for researchers engaged with health and social welfare work this is a primary responsibility. This is turn reinforces the issues around creating and analysing knowledges, and ensuring that this all happens in the public domain. As a direct consequence of their roles in any community, professional practitioners find themselves at the heart of issues about researching, theorising and the application of such knowledge to do good rather than harm. It follows that those who design and deliver educational programmes for professional practice are themselves involved in such concerns about ethics and empathy. Of course the increasing diversity of providers will make everyone involved more aware of the potential dangers of good governance being a victim here. One way of minimising potential problems here is for practitioners to take on these responsibilities themselves. An example of this issue that links to action research is discussed by Ortrun Zuber-Skerritt in his book *Action Research in Higher Education* where he is critical of the shallow understanding of the relationship between research being done by academics and so on, noted by and acted on by practitioners, and changes in a community being initiated for better or for worse. He argues that within higher education itself practitioners should be more aware of the dialectical relation between doing and using research and taking action in a community.

> Action research reflects this dialectic (i.e. action and research are like two sides of a coin). Action and practical experience may be the foundations of educational research, and research may inform practice and lead to action.
> (Zuber-Skerritt 1992 p. 11)

He goes on to identify his preferred way of describing action research, which neatly underlines the issues I have raised. He says action research is:

- Critical (self-critical) collaborative enquiry by
- Reflective practitioners being

- Accountable and making the results of their enquiry public,
- Self-evaluating their practice and engaged in
- Participative problem-solving and continuing professional development.

(Zuber-Skerritt 1992 p. 15)

My argument here is that action research can contribute to the well-being and sustainability of a community, and is more likely to engender positive outcomes and community building, if all stakeholders are actively engaged in the processes.

There are important links in the community-building goal with social capital (which I address from other angles later in this essay), for example differentiating between 'bonding' social capital, the infrastructure of the community/culture, or perhaps the 'bridging' social capital where more focus is placed on developing links with people 'beyond' a tight-knit group. In my experience this is also where 'outsiders', for example social scientists working in a local-ish university initiate, respond to and collaborate and cooperate with a group of community builders who wish to use a research process to bring about change. As I have already said in this essay the value orientation of social scientists makes the exploring/solving/explaining and understanding situation pregnant with possibilities.

In terms of our ethical stance in relation to research and educating we could do worse than to consider the United Nations Convention on the Rights of the Child (UNCRC, ratified in the UK in 1991). The UNCRC has at its heart the 3Ps: protection, provision and participation, and in our everyday role performance we could all benefit from reflecting upon and embracing these principles.

These are democratic cultural values and ideals, much needed in an era of competitive individualism. (See Astley 2008 for a further discussion of these issues, where a consideration of just how much of a struggle is involved to keep the language of alternatives to the status quo on our daily agendas.)

As educators we can use our role to introduce people to ideas and issues they did not realise they needed to know about, and are unlikely to get from anyone else, and certainly not from the current crop of 'managers' in further and higher education.

Despite the many differences in the vocational group particularities of the professions, there is always a fundamental social scientific focus to most educators' input into courses. My own practice has been so, bringing a common core of theoretically informed knowledge across programmes. Regardless of the specific setting I remain the same sociologist concerned with social policy contexts, carrying my tried and trusted experientially honed intellectual toolbox with me. Of course I remain sensitive to the cultural variations amongst and between these vocational group settings, but my perspective as a critical theorist placing the highest value upon a liberating education is a constant factor in my role and approach to curriculum development, design and delivery.

I mentioned the Open University earlier, and I would like to introduce here a brief case study to emphasise some of the good work done on roles.

***

Case study

Some colleagues in the Open University have gone a fair way to exploring key issues around the role of educators. For example the work of Maggie Coats and Jo Tait comes to mind, as they tackled the issue of helping distance learning tutors develop their practice from one of competence to excellence. Jo Tait's research identified three strands in the effective development of tutors.

1 An appreciation of their complex professional lives,
2 Their diverse work and learning activities, and
3 The range of motivations they may have for seeking development opportunities.

She also suggested that part-time tutors (the norm in the OU) could well benefit their own practice, and the academic community around them, by actively engaging in collective endeavours. I was briefly involved in complementary work that Coats and Tait conducted which amounted to engaging tutors emancipatory action research, which addressed Tait's criteria above, and in my judgement showed tutors a reason for, and route to, virtuous practice. (See OU publication Cobe Web Issue 5 May 2004 from the Centre for Qutcomes-based Education at the OU.)

***

One key dimension to the complexity of the tutors' role is the increasing use of computer-mediated communication. The Open University led the way as a virtual education provider, and organisation in general, and was certainly engaged in a series of (often unacknowledged) experiments focused on group communication and cooperative learning. What has often been referred to as 'nattering on the net' seems an ideal modus operandi for a distanced learning institution par excellence. However, the social systems in place at that time (still are?) were not conducive to working in those ways. There are many ways in which IT can be used to assist tutors in their role and in the support of students.

Much more co-ordinated information about the 'faculty', the curriculum offer and its development, about research, and appropriate information on tutors and students could, literally, be at our fingertips.

In an era of increasing IT use and its domination of everyday communications, educators need to look very carefully at the implications of less face-to-face interaction for all involved.

To conclude this chapter I only wish to add that our roles as educators are familiar to us over time, and yet open to all manner of changes, for example as discussed in the Open University case study above. Whether these changes in our roles come from ourselves as professional practitioners, social scientists or not, or from other stakeholders, we need to keep our focus clearly on the service to aim and claim to provide. I take up this altruistic dimension to our practice in the next chapter on communities, in which we all live and work.

Chapter 5

The community, providers and professional practice

The needs that professional practitioners routinely respond to exist within a particular locality. Those localities are invariably referred to as communities, as are the people who live in that place. Indeed particular culture groups – denoting ethnicity, religion and so on – who inhabit a specific place are often described as a, or the, community. They are apparently identifiable in several ways, including symbolically as an indigenous social group. Increasingly politicians and policy makers in general assume that care will happen in, or even be provided by, the community. So, in short, there are at any one time in contemporary British society a multitude of assumptions being made about the existence of all these 'communities' even if there is often little evidence to suggest that they are real other than in people's imaginations. People in society still do put positive meanings upon these various understandings of community regardless of a significant lack of either evidence or understanding. A person's sense of self, their identity, may well draw on these reservoirs of meanings. This can of course be extended to a sense of identity amongst a culture group to which individuals subscribe. This takes us into the conceptual area of 'the social bond' mentioned earlier with links to a discussion about social capital. However, it is clear that many people are often caught between what the sociologist Karl Mannheim called 'ideology and utopia', the fictions that crowd into our politically and media-driven information-gathering scope on the one hand, and the wish dreams that we still hope to realise one day. There is also a lingering sense amongst many people that the old style inclusive and supportive community did exist, but in the past, in some 'golden age' of symbolically significant social relations where people were within *the* community which has just slipped away, out of grasp if not out of mind. This nostalgia about, and melancholy for, the loss of inclusiveness have been re-fed into a rhetoric of bringing 'power to the people'. Such ideas of a genuinely pluralistic, democratic and open society have been regularly dusted off by politicians in the name of 'localism and the big society' (Cameron), active citizenship (Hurd/Thatcher), communitarianism (Blair) and so on. These adventures in community building have usually been top-down exercises, and existed in a parallel world from the many bottom-up attempts by people of a particular locality to devise an infrastructure where the management of social arrangements responds more to their needs. There is also a tendency towards a 'self-help' model of meeting needs here linked with a belief from some that we are all, or can be, altruistic citizens. How much reciprocity is there in contemporary society? Can we talk of a lively civil society?

Of course there is an extensive sociological literature on the quest for 'the good society' which draws on and documents issues like those raised by Mannheim in the 1930s. Even the growth in 'nostalgia TV' feeds this sense of loss, without offering any serious route map for people to follow out of this often emotional bind, and it seems only adds to a sense of loss and the onset of melancholy. These variants on social myth-making do not hold back the everyday work of activists and practitioners in the community, but the lack of a collective identity–led focus on alternatives does act as a brake on taking on conservative forces on both sides of the provider/client line.

In this chapter I intend to address these paradoxes and discuss from my sociological perspective issues pertinent to the meaning of 'community':

1 What is it?
2 What does it mean to people, for example to any group of local citizens, or to professionals?
3 Is there a significant distinction between what professionals, and the organisations that employ them, believe to be 'the community' and the understandings of other people including people with certain welfare needs? Community as fact and/or ideal type?
4 And given this what are the prospects for a feasible 'care in the community'?
5 And, what are the implications of the questions, and some answers, to the education of professional practitioners, and the sense that those practitioners have their role to play in all of this?

In many ways these questions return me an earlier question raised this essay, namely 'what is the nature of the social upon which we are working?' and most certainly promotes the opportunity to talk about governance.

My aim is to explore a range of explanations for what community is, and how these conceptions of community are then employed to prepare practitioners of various kinds to 'enter' these locations, in mind and in physical reality, to make provisions of diverse kinds. This will take me into a discussion of public service and the role of professionals in both identifying and meeting needs. One key issue to explore is: exactly who is doing the identifying and providing, and which organisations are they working for and within? It is clear that an increasing amount of social welfare provisions that are devised and deployed to meet needs in a community come from non-statutory organisations. This increasing role of the voluntary and charitable sector, and by social enterprises, raises important questions about the preparation for professional practice.

It is a sociological truism that social change is constant, if uneven, and that its management will always remain a major issue for all citizens regardless of their situation. As already suggested, one aspect of such change is a constant focus on the inter-relation between social issues and problems, and personal troubles. For example how does any individual deal with everyday needs, and plan for the future? Social welfare need is a broad concept, but we all have to think about how to seek provision to meet those needs whatever they are. Do we enter into the marketplace and purchase what we need: a house, a care treatment, a knee replacement, an education for our children and so on? Or do we rely on provisions from the state apparatus, or from a community-based voluntary sector provider, or from charity? How do individuals, and small social groups like a family, make life less precarious, less chaotic? To whom within the 'community' do we turn for service, support and succour? What societal and personal benchmarks exist for effectively differentiating between needs and wants, and how does any 'community' play a role in both defining these benchmarks and taking action to achieve this? One issue that has become much more of a focus of attention in recent years has been the probable/possible health and well-being benefits to people of both living in a caring community with plentiful high-quality provisions, and believing that to be so, foremost among them a sense of security.

Many of these issues were addressed by Richard Wilkinson and Kate Pickett in their 2009 book *The Spirit Level: Why more equal societies almost always do better.* Both authors are epidemiologists writing as public health researchers, and drawing on very comparative and data rich evidence, the very backbone of debates about why in a society of plenty, many individuals and especially families can no longer cope.

It is also the case that people with social and welfare needs, and even many providers in a particular locality, are confused about exactly who is responsible, or even accountable? Since the 1970s there has been a steady process of the centralisation of state power over policy and 'national' standards, and criteria in terms of provisions, while at the same time adopting an aggressive de-centralisation of providers. We have witnessed an increase in quangos, next-steps agencies e.g. the Highways Britain agency, and extensive recourse to outsourcing by all sections of government. The National Health Service (NHS) and local authorities have journeyed extensively down the outsourcing road, finding providers from across a multitude of sectors including the use of explicitly profit-making organisations, usually via tendering. There has also been the public finance initiative (PFI), used extensively by Blair and Brown after 1997. One personal example of my being drawn into this new world of running the state apparatus was being asked to join my local New Deal steering group. New Deal was created by Gordon Brown to deal with the extensive problem of youth unemployment and devising ways to adequately prepare young people to move into work. This policy-to-practice initiative ran parallel with extensive research on social deprivation run over several years from the Cabinet Office at the heart of government (the 'National Strategy for Neighbourhood Renewal' with distinct echoes of the Community Development Programme of the 1970s). The money to pay for this two-year policy initiative came from a windfall levy on the recently privatised utility companies. Even at this early stage of the New Labour management of the state apparatus they were keen to involve the private sector, and in my area the New Deal scheme was managed by the Employment Service (as it was still called at the end of the 1990s), but they were required from the outset to work with a private sector involvement pilot scheme. In this instance the skills training and employment company A4E (Action for Employment) was a co-provider. I do not have the space here to detail the many problems that arose from A4E's involvement, but I can say that the delivery of services allocated to them, and allegedly supervised by the local government including the steering group, was beset with problems.

There were and are many similar issues associated with the Blair/Brown enthusiasms for 'small government', and even governance. For example one such early policy and practice initiative were Primary Care Trusts (PCT), created to bring the essential care and services of primary care to *the* locality. People's needs could be addressed and managed, and with a greater likelihood of good governance. Alongside the PCTs, the Commission for Health Improvement (CHI) was devised and set in motion to enable the peer review of health and social welfare staff, including non-executive directors. The hope here was to keep governance high on the agenda of both providers and recipients in any locality. I did some work for them as a lay reviewer alongside health care professionals, and what an interesting process it was. However, as we all now know, the small-scale approach to PCTs was soon abandoned, and CHI was merged into an ever-expanding inspectoral regime.

Another aspect of social change is the manner in which the state, in the process of developing social policy and practice responses to situations, re-considers the actual roles of practitioners 'at the chalk face'. The naming and re-naming of practice roles has been commonplace over many years, but has certainly been so in the last thirty years. The state has often considered collapsing together the role of certain practitioners in order to convey to potential clients the actual purpose of the role holders. For example a few years ago there was an enthusiasm in the state apparatus to reform the children's workforce, and one aspect of this was to create a new role of personal adviser, primarily but not exclusively in the Connexions (careers) service. The existing practitioners 'invited' to take on this role came from careers or youth and community backgrounds in the main. Research done by Billie Oliver in Bristol suggested that while practitioners were concerned about a loss of professional identity, and even status, they were more concerned that clients, or other professionals, would not easily identify with them as 'proper' and knowledgeable professionals if there were no history to that role (Oliver 2005).

So roles and interventions into people lives, and the perception of social change, can be confusing, misleading, especially so with hindsight. Even in the good old days of abundance in the 1960s, life was not 'swinging' for everyone. These were the days when 'Cathy' did not go home (Sandford and Loach 1966), when Coates and Silburn (1970) cast a light on the slums of Nottingham, and when Peter Townsend (1979) needed to remind us of the relative poverty of many amongst the plenty, where even the conventional utilitarian policies and provision of welfare were seen to be inadequate. Since then all the major political parties have abandoned the post-1945 commitment to a social democratically led 'welfare state' and increasingly left a liberated citizenry to their freedom and choice in a society increasingly dominated by a global capitalist consumer culture, where want dominates need, where services have replaced manufacturing, and social mobility has stalled.

Those contradictory and contrary pasts have tended to be buried by a collective political and media-led amnesia, but, inevitably have now come back to haunt us all like Marley's ghost.

Before I take this argument into a discussion about the development of contemporary social welfare provision, and the providers, I should return to the generic issue of 'community'.

Community is a contested concept, and like statistics can be a musical instrument upon which any number of different tunes can be played. For example, as suggested above most definitions of community incorporate the influence of culture groups, which can amongst other things be based on ethnicity, nationality, social class, religion, gender and region. All these factors are, and can be, relevant to an understanding of what a community is in the eyes of some people, and an understanding of why and how it works. The values of such culture groups are crucial to the ways in which both the assessment of need and responses are made. The real-life experiences of diverse cultural groups, like communities in general, have created an extensive fund of knowledge, the know-how for coping with the 'snakes and ladders' struggles of everyday life. This know-how, described by Raymond Williams (1965) as a 'structure of feeling', invariably contributes to practical solutions for evidently unmet needs. I see the resurgence of this approach to welfare needs being reasserted in the community building and infrastructure connectivity of bottom-up community-based providers, of which I will say more later. I shall also return to the relevance of social capital as a concept which helps to develop our understanding of this community know-how.

For many years now most people's understanding of welfare provision has focused on the role of the state, nationally and locally. In the domains of health, care, benefits and services (like education) citizens have been assured that the state would take the lead in addressing need, for both collectives and individuals. This is not to overlook the almost continuous role of charities and the wider voluntary sector as both a necessary supplement to state provisions, and a reflection of civil society at work, with its pluralistic ideology and tireless support from 'the great and the good'.

When we talk about the public sector this is what we are considering. However, how often do we pause to ask ourselves: what is the nature of this 'public' that we are making assumptions about? In recent years the shifts in political ideology has meant that the vanguard role of the state has been increasingly questioned, moving from a social democratic 'big state' approach to the neo-liberal 'small-state' one. One of the direct consequences of these changes is that the politics of selfishness has been increasingly evident. We have also witnessed the often paradoxical 'withdrawal' of the state as an aspect of the 'small-state' ideology, while there has been a concentration of state power in Whitehall. There has also been an attempt to divest various Secretaries of State of their direct oversight and responsibility for the daily organisation of provisions. The shift to 'next-step' agencies of government, an allegedly semi-autonomous management of need, has happened, while leaving the state as the impresario, the guiding hand of policy and provision.

While all these changes have been taking place practitioners of many kinds who have staffed the various state organisations at a 'local' level have carried on delivering care, benefits and services. It has usually been the case that these people, all citizens and service users themselves of course, have been left out of policy-making discussions. The state has known best, but usually advised so by big money interests. What local practitioners know and do remains a crucial aspect of the social capital, the know-how, of any community, whether that is the district nurse, the primary school teacher, the GP, or the youth and community worker. It is also the case that the education and training of most practitioners has been greatly influenced by the social scientific conceptualisation of contexts already discussed in this essay. To see the fragmentation of these key community role holders, and witness the loss of expertise and input, poses serious questions about the ability of any community to both assess and respond to need.

I should also add that to even be discussing a community focus is relatively new, again. For many years in the mid-twentieth century the dominant planning focus outside of Whitehall were regional authorities of one form or another. Now, the majority of those have all gone, and we are back to 'community', locality, 'localism', but still of course with the constant presence of *the* local authority somewhere in the background. (For more on this see John Urry in Harloe et al 1990.)

As already argued above these many changes in planning, policy making, provision and providers, these shifts to different models of assessment and response, do emphasise the importance of networks and networking. Building the capacity of any community to respond to growing or different need is clearly crucial whoever does it. Who is to give leadership, co-ordinate and facilitate the 'people power' that is both latent and expressed in any community? One aspect of the liberal-democratic society in which we still live has been a widely held belief that our political representatives were/are the essential conduits through which our values, desires and grasp of needs are directed in order to meet those societal and individual needs – civil society in action. However, it is very evident that many, even most people in society today have lost faith in the political 'class' to fulfil this key role. In most communities people look around them and see a democratic deficit at the heart of decision making and action taking. It is in part because of this current malaise that citizens have been calling for a re-assessment of the role of the community, even their community, in everyday life.

In a way we are returned to questions about values, and about choices, which arise every day in policy making and the eventual delivery of care, benefits and services. How do organisations and the professional practitioners who have a role in social welfare–oriented organisations make choices? In an article in *The Guardian* about caring for children Alison Gopnik cited the political philosopher Isaiah Berlin on 'value pluralism'. Berlin argued that any society has a multiplicity of values about welfare, and those values are often incompatible, and therefore advocates of this or that priority 'compete' for special consideration. Gopnik employs Berlin to focus on the key societal issue of caring for children:

> In fact, being a parent is valuable precisely because it is so unlike goal-directed work. Caring for a child involves a deep recognition of the individuality and autonomy, the irreducible complexity and value of another unique, irreplaceable human being. That makes it worthwhile in itself.
>
> (Gopnik 2007)

I would suggest that most professional practitioners are faced with such value judgements most days of their working lives.

However, just because a non-state provider organises and delivers care or benefits is no guarantee that the dominant profit-making motive of the commercial sector will not take centre stage. The significant 'privatisation' of health and social welfare provision has already seen a transfer of this provision to explicitly commercial organisations, for example in housing, health, education and non-statutory social services. The voluntary and charity sector continues to be a major, indeed increasing, provider, but often with significant changes in mission and management ethos. In the last thirty years or so many, particularly national voluntary sector, organisations have embraced a set of management values that have mirrored the style promoted in the Thatcher years. Her governments had a penchant for challenging professional orthodoxies, and particularly so in the public sector. She saw professionals operating a system of restrictive practices. Thatcher and the neo-liberal caucus in general disliked (that is probably too mild) the public sector, and could see no good reason why creating and managing provisions to meet all human needs should not be turned into a profit-making opportunity. One by one she took on the professions, working within in the public sector seeking to impose her views on management style and operating values.

There were major changes made across the public sector, and the effects of that still ripple through the sector. However, this dominant set of values also infiltrated the voluntary/charity sector and several organisations changed their management ethos, often deciding to employ senior staff from the 'business world' who did not have a public sector professional background and were set on making organisational changes that put hierarchical values and money making above all else. I have discussed this issue elsewhere in this essay and mentioned the frequent shift from a fundamental focus on transformational relationships between staff and with clients, to a transactional approach.

The relationships here between state-sponsored quangos, trustees and the funds provided by the general public (including lottery funding) need much clearer examination; more transparency would help accountability and good governance.

I mention this issue here because there are several places in my discussion where the shifts in provision are discussed, and we should try to avoid naivety in an assessment of these significant changes.

Whoever the providers of care, benefits and services may be, they will be addressing a range of social issues as well as personal problems. The rise in concerns about *public* health in the UK now is an example.

In the mid-1990s I was asked to design and deliver a course on 'Social Issues' for a university based M.A. in Social Work. As was usual I produced both a course handbook and a course reader for the students and colleagues. Given what I have already said about community and contexts to the work of practitioners, this extract from the course handbook is relevant.

***

Why social issues?

As a practitioner I need to be able to understand the context in which service users live and within which they and I together have to address whatever needs or problems have brought them, voluntarily or involuntarily, to my agency. Parts of that context will be political, parts will be geographical, and parts will be socially defined. It is those parts which this course will seek to explore. We will be considering what society in this country, at this point in history, defines as "acceptable" or as "problematic" and how such definitions impact on both those within the "problem" groups and those outside. In addition we will consider the expectations which society has of its individual members, particularly in terms of age and gender, and how these expectations influence our perceptions of ourselves and others and the consequences, both positive and negative, of our fulfilling, failing or deliberately rejecting these socially prescribed "norms". Inevitably this will lead us into some consideration of how society seeks to perpetuate and indeed enforce these expectations and the part social work may play in such processes. Throughout we will attempt to make links with practice examples and use these to explore the implications of our developing understanding for different approaches to the social work task.

***

One of the core themes addressed throughout the course was the concept of 'life course', which was one of 'issues' to be addressed and which I signalled by putting them on to the cover of both the handbook and the reader. The others were: sexism, opportunity, illness, family, low incomes, inequality, racism, disadvantage, unemployment and deviance. The course text was *Social Change and the Life Course* edited by Gaynor Cohen, a 1987 Tavistock publication, and I felt that this approach, focusing on a life course theme, helped students to see the dialectic nature of their response, engagement and reflection.

A key issue to be addressed here is the nature of those organisations that are the providers of social welfare. As the state, at both national and local levels, has 'withdrawn' from a good deal of actual day-to-day provision, other forms of organisation have taken over. A good number of private sector organisations have been given contracts to deliver care, benefits and services. Successive neo-liberal governments, regardless of political party, have an aversion to public sector providers, and wish to create a climate that encourages people to make private profit from the delivery of public provision. One of the many negative outcomes of this process of the increased commodification of needs and responses to it has been changes in the terms of pay and conditions for staff, not always in the short term when the transfers are taking place, but certainly in the long term – out of sight and out of mind. Governments have also made policy decisions to recruit organisations from the voluntary sector (what state officials like to call the third sector) on a series of contracts. The resultant picture is a very mixed one, with many voluntary sector organisations taking on work as a valued client-specific response to needs. Many voluntary organisations and charities have changed their ethos to a distinctly market and managerialist one, eager to embrace the neo-liberal model on offer. This has cast a long shadow over parts of the voluntary sector. One of the reasons for such outcomes is that organisations are people, and the sets of relationships both within the organisation and with other people have often changed. Internal relations contribute a large part of what constitutes any organisation's culture; it is a way of life, replete of course with considerable symbolism. Having an understanding of the relation that members of an organisation have with 'outsiders' is also crucial to making sense of the nature of that organisation, understanding motives and so on. A good deal of any organisation's symbolic activity is invested into creating and maintaining boundaries, thereby creating outsiders, separating this 'community' from that. When we think about an organisation, especially so one we are regularly involved with, it is important to remind ourselves that we are not just thinking about what people do, role performance and the like, but what meanings they and we place upon what they do as part of the inter-relation. I am not going to argue here that it is only professional practitioners in the 'people industries' that take issue with their practice, are reflective about it. Nor do I want to suggest that it is easy for us to focus clearly on the complexities that make up practice relations within providers. But deal with these complexities we must, because in an era of constant and far-reaching change, including the fragmentation of conventional modes of practice within organisational contexts, roles and their value-laden criteria are shifting.

To return now to the beginning of this chapter, I posed a number of questions about community, the first of which was 'what is it?' I shall now offer a brief explanation.

Sociologists have been studying 'community' social relations, and the social relations that might be seen to make a community, for over a century. This is largely because in the industrialising societies of the late eighteenth and early nineteenth centuries the forces causing migrations between rural and urban, in both directions, had been such a potent issue. For example a good deal was written in the early twentieth century on the social and cultural disequilibrium caused by these migrations, especially so the dramatic population growth of industrialising towns, most of which did not have any parliamentary representation until after the Reform Act of 1832. The impact of migration is still felt on a worldwide scale, and continues to produce divergent attitudes towards migration amongst the UK population.

Most writing on community has emphasised the importance of shared values linked to local cultures encompassing shared work and everyday life experiences. On the one hand these 'new' forms of social relations were often seen as an emergence of 'the mass society', a phenomenon that usually had negative connotations culturally and politically if not economically. On the other hand, the creation of an increasingly self-consciousness working class and a concomitant growing middle class servicing the state apparatus encouraged people to focus on the developing cultural homogeneity within classes.

It should be added that discussions about 'community', both real and imagined, all with their assertive creation myths, are close to arguments about nation and nationhood. It is clear that accounts of nation building are ancient, and that the link between the development of assertive religion and national identity are very close. Indeed many nations, including those countries that make up the 'United Kingdom', were driven by the commitment to *the* dominant religion of the era. The 'relationship' between aggressively protestant Britain and popish Ireland is a good example. It is also the case that 'God' appears to be on the side of the assertors, often the victors. So when considering the merits, or otherwise, of any one 'community', we should consider the inter-relation between this or that community and the nation of which it is allegedly a constituent part.

Sociologists became particularly interested in the development of the shared bonds that could be seen to make up a 'community' involving the fraternal sharing and caring aspects of people with a shared locality – physical and symbolic space – and day-to-day life experience. German sociologists call this *gemeinschaft*, a holistic sense of belonging. As I have already suggested, Raymond Williams' conceptualisation of a 'structure of feeling' helps in the sense that these values do have meaning and give coherence to peoples understanding of their identity. Richard Hoggart (1957) also discussed how working-class communities had been 'invaded' by, for example, Americanised pop and consumer cultures. Communal support exists on the basis of shared understandings of circumstances and needs, which required a collective approach to meeting those needs. A 'home-grown' resilience develops to deal with adversity, and altruism is effectively built into the fabric of everyday life. Trades unions, insurance schemes, local charities, religious organisations and so on all provided a measure of social welfare provision to enable people to deal with frequently and periodically precarious lives.

It is the case that most 'communities', regardless of who exactly designates them as such, are indeed made up of people who share in good deal in common culturally, including a history of place. We should of course be cautious about such attributions for the existence of this or that community, and keep asking questions about why these or those people choose to identify with a particular collective. What is more commonplace then are 'communities-in-the-mind' (Pahl 2005). Insiders and outsiders, the 'other', observers near and far may like to believe that such and such is true of certain people in certain circumstances. Behaviours will certainly be attributed to aspects of community life, often as it suits people to do so. Perhaps the big question is always going to be whether being a member of a physical and social community (a particular group of people in a certain locality at a specific time) will have a raised cultural and political consciousness that engenders a social solidarity that will impel them to take action collectively? Be prepared for threats to the enclave is the watchword.

A key issue that has regularly been discussed is that of forms of social stratification in any particular place. This might be strata based on social class (usually occupation related) or ethnicity or gender or religion, or a combination of these in some complex historical way. Whatever the mix, there will be symbolic evidence of hierarchies in that community, and this will certainly affect most aspects of social relations in everyday life. For example deference has a very physical presence in most communities.

Access to employment, where a 'closed shop' based on some social differential, would be another such example. Schooling and education would be another example, and Gary Easthope is typical of sociologists who have addressed this issue in his 1975 study, 'Community, hierarchy, and open education'. Easthope suggests that the idea of community can be looked at fundamentally in terms of the relationship and conflict between individual freedom and collective organisation and centralisation in people's lives. Easthope argues that schools are inherently hierarchical, and are but one key aspect of the form contemporary societies take.

Basil Bernstein in his pioneering work in the 1960s and 70s argued that our place within a community does affect educational outcomes:

> Working class community rests on the division of labour in society. The working class family has few, if any, financial assets. The man has nothing to live from except his labour and his skill. His job is unlikely ever to give him enough money to build up capital, let alone employ others. This . . . economic division, whatever else it does, pulls the working class together as a group, and as a self-conscious group.
>
> (Quoted in Jackson 1972 p. 164–165. But see also Bernstein 1970)

Of course since Bernstein and Jackson reflected on this, many sections of the 'working class' have become members of the 'property-owning democracy'. However, the educational success rate of successive children from this section of the population remains stubbornly low, suggesting that differential educational achievement is still creating a lack of social mobility. This in turn is increasingly felt as another aspect of unfairness, that key barometer guide to rising frustrations, even anger.

Several researchers work came closely together at this time, with for example debates about the role of 'cultural capital' being prominent. Amongst other sociologists, Pierre Bourdieu (1977) argued that families tended to pass on to their children a cultural know-how, a set of expectations, a guide to navigating schooling and work opportunities. Not surprisingly he suggested that middle-class families were in the best position in society to promote, even sponsor their children's progress in a schooling and education system reflecting middle-class life, values and aspirations. In the UK Basil Bernstein, already cited, agreed that schools were 'cultural repeaters'. He said much the same with a particular focus on language acquisition and usage. He argued that despite various reforms made to the schooling system, in general, and in response to the concerns of the state about falling social mobility, 'Education cannot compensate for society' (1970) was the most likely outcome for working-class children. Even stalwart advocates of these various minor reforms to the schooling system began to have second thoughts during the 1970s, A. H. Halsey's 1980 study 'Origins and Destination' being one such example.

In one way or another, the form and fabric of a community will have a bearing on the nature of need and of the nature of care in *that* community. As suggested above these factors may well change over time, but the pattern is uneven. This is often referred to as the 'post code lottery' of care, benefits and services. Interventions into communities to meet needs, to provide care, may reflect the diversity of that place and those people, but this will depend on the nature of the providers, the assessment and analysis of a knowledge base that indicates what action should be taken, the role of various stakeholders, and the combination of policy and practice.

One such example of these issues is discussed here in a case study written for this book by Tim Tod.

***

'Youth work in school' by Tim Tod

As someone who has only ever supported and delivered youth work from community settings a phone call from the head of Plymouth School of Creative Arts asking if I wanted to base myself within his leadership space was a really interesting proposition. Especially so as I had recently decided to step down as

CEO of Young Devon, a major voluntary sector organisation, after 20 years in that role.

As I settled into my new environment I began the process of examining how I could work alongside staff, young people and parents. Firstly, I needed to explore how youth work could truly flourish in an environment that at least on the surface operated very differently:

> Youth work is based on relationships voluntarily entered into, schooling, for the most part, is compulsory. Youth work is relationship focused, schooling is curriculum-driven.
>
> Mark K. Smith (1996) *Local Education*

So, in practice I decided that the best way to examine this in more detail was by linking the school into a youth work organisation (UK Youth) and one of its projects – The Big Music Project (TBMP), and see what happened along the way.

Firstly, let me tell you about Plymouth School of Creative Arts and UK Youth TBMP:

> Plymouth School of Creative Arts is a newly established 3–16 mainstream, city center all-through free school sponsored by Plymouth College of Art. One of the main reasons for gaining government support in establishing this school in the first place was the acknowledgement that it could act as a beacon of inspiration and hope, within a community that historically has been a community struggling to transform itself with many of its residents living with multiple and significant disadvantages.

The stark reality as highlighted in the Plymouth child poverty report (GOV data) is that the school is situated in the St Peters and Waterfront ward where 45% of families in are living in poverty, compared to the Plymouth average, 27%. It also scores 1 as one of poorest wards in the whole country according to the Index of Multiple Deprivation (IMD) decile (where 1 = most deprived 10% of LSOAs [lower-layer super output areas]). The Red House, where the school is based, was formally opened by Sir Nicholas Serota, Director of Tate, in the Autumn 2015.

The Big Music Project is a partnership between UK Youth (the largest national charity for the youth sector) and Global Radio (who own Capital, Capital XTRA, Radio X and others), funded by The Big Lottery Fund, creating UK-wide development opportunities to 14–19-year-olds. The consortium brings the full weight of the music industry into play, using the reach of Global Radio and their Capital Brand and the grassroots networks of UK Youth to connect with millions of young people around the UK through their love of music.

The Big Music Project equips young people with the skills and knowledge they need to create a successful future in the music and creative industries, as well as building their general employability skills. We work with the biggest brands in music, breaking barriers and creating and highlighting the best opportunities for young people across the UK, running everything from training schemes and careers advice to work experience, internships and hands-on skills-building projects. We inspire young people by showcasing the range of roles within the music and creative industries, motivating them to launch their careers.

It was clear from the outset that as this project needed to take place out of school time and needed to involve young people who wanted to become involved, and that we needed to have a member of the workforce who was willing and able to lead the project (with me in support). So I was delighted to have it agreed that Dan Smith (producer, archivist and media teacher) would lead the project.

When it came to identifying young people we used the existing communication channels already established within the school to promote the project and within a couple of weeks a group of 15 young people had come forwards to be involved.

Because the project involved two trips to London (one at the beginning of the project and one at the end) it was interesting to observe how that was managed and their attention to detail. Every aspect of the trip was mapped out fully with detailed risk assessments carried out on all elements, with an excellent authorisation and escalation process in place for those accompanying the young people. The thoroughness of the process was tremendous and many youth work organisations could learn loads from this depth and breadth of best practice.

One aspect of the project that I was really keen to observe was the process of selection of interested young people, as expressions of interest outnumbered the number of places available for the London trip. There was a determination to ensure that not all of the group only comprised of their already highly focused and motivated young people but also a few with latent potential as well.

Once the core group was established and after their attendance at the first London training event the young people soon became focused on putting on their public-facing event – a silent disco with additional input from a mix of school performers. They also wanted to see if they could generate enough money so that it could be reinvested into the project so that the initial project would live on beyond its one year of funding.

Their ambition was tremendous and their commitment to the project was really excellent to observe. The event was a great success and although it did not take sufficient money to cover all costs it was also seen as a really worthwhile venture from the school's point of view, and it has acted as a springboard for the school to fully support the group on into next year. Feedback from staff, students and their parents has been very positive; and we now have many more young people coming forward with ideas and enthusiasm keen to become involved in next year's project.

My initial anxiety about how the school staff would work alongside a youth work project was completely allayed. Their enthusiasm and willingness to support the aims of the project and its informal education approach was great to observe. Over 10 staff volunteered to help with the silent disco, which together with the group of young organisers was enjoyed by all and raised over £500 towards their next youth-led event.

I was able to bring my many years of practice experience to this community role, and share with a wide range of stakeholders an opportunity that required an imaginative use of all of our resources.

***

One of the contemporary paradoxes here for professions in general and conventionally educated and experienced professional practitioners in particular are the ways in which they are seen as both knowledgeable and central to assessing and meeting need, and yet increasingly marginalised as social welfare provision fragments. I have addressed this issue of social change within several different contexts in this essay, and as we are now faced with the most significant public health crisis for decades, these issues of roles and engagement are coming to a head. The longer-term consequences of developing notions of the 'social model' of health and social care have already posed serious questions about traditional modes of intervention into people's everyday lives. If we are to say that interventions of all kinds – health, education, social care, housing and so on – should be focused much more on preventative work, action rather than re-action, this does argue for a radical re-casting of roles and relationships. The more enlightened social reformers in the late nineteenth and early twentieth centuries knew this. They put public health onto the social and political agenda precisely because of an understanding that without the fundamental building blocks of an appropriate standard of living, a reasonable quality of life was not possible. Early twentieth-century welfare reforms like old age pensions led directly to questions about 'insurance' against the precarious nature of life in an industrial capitalist society, which led directly to the Beveridge Report of 1942 and his focus on the 'five evils': want, disease, ignorance, squalor and idleness.

In one way or another; in schools, in health and social care, in youth services, employment services and so on, the state has nudged pluralist thinking about social welfare in the direction of more self-help solutions and greater use made of semi-professional role holders. So alongside teaching assistants, and their equivalents in other sectors, we have seen more volunteers. We have also seen an increase in interest in 'social prescribing':

> Social prescribing involves empowering individuals to improve their health and well-being and social welfare by connecting them to non-medical and community support services. It is an innovative and growing movement, with the potential to reduce the financial burden on the NHS and particularly on primary care.
>
> (Kings Fund 2017)

In Exeter close to where I live there has been a community-based provision development of considerable interest. Some years ago Exeter Community Volunteer Services (CVS) worked with local stakeholders and professionals to create CoLab, a multi-practice experience for the most needy individuals and vulnerable groups in the city. The very fabric of their building was re-designed to create an entry and potentially, an exit point, for a range of clients whose needs could be met by a range of professional practitioners from GPs to counsellors, to housing agencies, to education provision. CoLab promotes itself as 'social innovation through partnership', and those city-wide partnerships have created 'Integrated Care Exeter' (ICE), an alliance of health and social welfare providers as an aspect of Wellbeing Exeter, a project that seeks to integrate social prescribing and community building. ICE is a strategic alliance of public, voluntary and community sectors organisations who work together to provide an infrastructure for designing and delivering improved ways to provide public services. 'The ICE model for population health and wellbeing sets out a model with a greater emphasis on early intervention and prevention' (Wellbeing Exeter Report 2017).

When teaching my social policy undergraduates in the 1980s and 90s I was made the point that the title National *Health* Service was from its 1948 origins something of a misnomer, and should in fact have been called the National Illness Service.

The ICE model emphasises the centrality of the/a community as an 'asset-based community development', where a community capacity building and volunteering approach draws on what sociologists call social capital, discussed within other contexts in this essay. There is a focus on the resources of the locality, even enhancing the considerable discussions of a decade ago about 'localism'. These local people and organisations-based developments are resources for hope in those communities, and seek to create a resilience to meet and overcome the increasingly precarious lives of people. The Wellbeing Report also emphasises that while professional practitioners and the predominantly statutory organisations that they work for are specialists and skilled in their roles they are often out of touch with the grass roots of everyday life. So a bottom-up model drawing on a widened range of analysts and providers, including many and diverse professionals, replaces a top-down approach. An important dimension of provision that this project has created is 'community connectors and builders', the actual practitioners who actually do the face-to-face work with clients.

> Key is the creation of a number of 'Community Connector' and 'Community Builder' roles to facilitate flow between local primary care and communities embeds this approach directly into what already exists but is now often unseen as an option for individuals. The roles guide people to identify and access 'what matters to them' for wellbeing and participation within their community, supporting positive lifestyle choices and preventative behaviours.
>
> (Wellbeing Exeter Report 2017)

The 'connectors' are a diverse body of practitioners often on 'secondment' from their host organisation, one aspect of the initial three-year duration of this project. There is also a move towards single competency and workforce development which seeks to focus on the core competencies that are desirable for working in this form of multi-agency and participative community-based health and social welfare infrastructure. The community professional development (CPD) demands that this approach to community working has produced is still being developed, but promises a more generic focus which in turn will certainly feed back into the already well-established professional practice course providers.

The political actions taken by the neo-liberals under Thatcher and beyond, and most certainly via the NHS and community care legislation in the late 1980s, signalled the closure of long-stay institutions of many kinds, with the resultant decanting of people into 'the community', coupled with a business-style management hierarchy imposed on the over-professionalised NHS. The current re-thinking and re-formulating of health and social welfare policy and practice in the widest sense can only be seen in this historical context.

The NHS Five Year Forward View of 2014 and several Wanless reports in the 2000s have argued that 'communities' and individuals must do more to assist the state in keeping us all (more) 'healthy' while avoiding through changes in attitude and behaviour the costly pitfalls of becoming ill. Who is to expect what from whom has certainly shifted.

And, as I have said elsewhere in this essay the enthusiasm for small-scale, non-state-apparatus, de-regulated and social model–type approaches to meeting social and individual needs (social issues and personal troubles) *may* create opportunities for a radical re-thinking of how to move on from decay, deficiency, disappointment and despair.

As already suggested one very significant aspect of welfare needs in relation to social change and viable and sustained provisions has been the growth of social enterprises (SEs). It has been argued that SEs can offer a locally based, more rapid response and 'can-do' attitude, a solution to both the ongoing assessment of needs, especially so from the client base, and devising ways and means to address that. Advocates of SEs have also argued that accountability and governance stand more chance of success in these locally responsive approaches. One of the advantages of SEs is that they can offer two major contributions to community life, creating jobs and economic growth, and thus through meeting needs have a positive social impact.

Social enterprises have been broadly defined as a trading organisation with primarily social purposes, and the growth of SEs has been so profound that both national and regional advisory and promotion bodies exist. Many local authorities, strapped for cash and devoid of the creative energy and wherewithal needed to take action on needs and provision, have increasingly turned to SEs. The whole notion of developing a parallel system of assessment and provision has been dubbed 'the shadow state', and although I like many observers have some reservations about motives, governance and accountability, the time is ripe for new strategies to emerge from the shadows into the light of public discourse. Social Enterprise UK is the 'umbrella' organisation for SEs, and seeks to raise awareness of, and advocate for SEs, as well as collect evidence about the development and effectiveness of SEs. SEUK produces an annual Impact Report which sets out the range and roles of SEs and assesses the 'social value' of these diverse approaches to meeting a very wide range of needs. The Public Services (Social Value) Act 2012 was an important step in the direction of everyone re-thinking their approach to provision, whatever their motives might be.

In Exeter, near to where I live in Devon, there are a number of SEs that have 'popped up' in recent years offering a very diverse range of care and services. There is also a local SE 'umbrella' group, ESSENCE, the aims of which are to connect, support and represent member SEs, especially so in strategic city planning and policy making.

One of the SE member organisations is Chime Social Enterprise, and below is a brief case study on Chime written by the managing director Jonathan Parsons.

***

Chime case study

Chime was formed on 1st May 2011.

We made use of 'Right to Request' legislation which allowed for a case to be made by a PCT Provider service to a PCT Board that a higher-quality, more efficient, cost-effective service could be provided if that service were to be a staff-owned social enterprise, directly contracted for the NHS service rather than employed by the NHS.

In the respect that we were a PCT service we were lucky. We had moved to PCT control in 2004 whilst physically staying located at the RDE (Royal Devon and Exeter hospital) and contracted by the RDE to provide services to ENT (ear, nose and throat). This was and still is a very unusual state of affairs as nearly all audiology services remain part of acute trusts. At the time in 2004 there was a long waiting list for hearing aids as digital had been introduced but there was no extra money or a target so it went out to two years. PCTs were newly formed and had their eye in community services. We had little say in the decision to transfer – RDE was sick of the number of complaint letters and the PCT was empire building. It worked for us though as it cemented our autonomous nature and separated us from ENT – where in a directorate we would always come second best to medics and the history had been in the not so distant past that we were mere 'workers' for ENT – doing their bidding, unable to talk about clinical results.

The second coincidence that helped us leading up to 2011 was that we started a conversation with the Exeter Academy for Deaf Education and were attracted to the idea of locating into an iconic new build that they were planning that put all sorts of linked services together on the new site. We were seen as an integral part of that new plan. The question was asked of us how were managed and budgeting worked and we got into a conversation about Social Enterprise.

At this time, we started to seed the idea with staff – circulating emails talking about the principles of social enterprise – what it would mean and how it might feel to own the company that we worked for. We gave examples. I would not say that many were overtly enthused at this stage but I guess some gave it thought.

When we got serious about Right to Request we met with Anne James, the CE of NHS Devon, and proposed the NHS spin-out supported by the Exeter Deaf Academy with their 150 years of experience. We got a green light to take things further and I think that was a critical part in the journey.

Then followed an intense six months. We had grant money to help us compile a business plan and weekly meetings with NHS Devon. In terms of a contract it was about coming together with an agreement on monetary value. This started from a very limited position. CCG: 'What exactly is it that you do? Have you got that written down anywhere?' to then saying 'What is your current budget? That is what you will get as a social enterprise'. Between us we had to unpick all the hidden and none explicit costs and extract a value. For example, human resources, IT, PCT management, capital replacement programmes, rent for locality clinics.

With staff we had many meetings – Audiology, Audiology with the Academy, with and without union reps present. The unions locally took a very practical stance – 'where can you see as a home that is in any way better?'. Nationally they were more cautious – 'will you not be at the mercy of bigger players after the initial contract?'.

In the end we did not take a vote as staff, we just accepted that it was the right thing to do. We needed Anne James a couple of times to intervene on our behalf with her board as they had no understanding of what was happening – wanted to get us to promise to buy all the services from the same place they had always been bought from. The fact that we were partnering at that stage with the Academy gave NHS Devon enough confidence in the governance arrangements that we would not immediately fail and I personally had to go through a formal interview with Anne.

When we formed in May 2011 the CE (chief executive) of the Academy was our chair and we utilised three of their non-executive directors on our board. We are a community interest company limited by guarantee – with at the time of spin-out, the guarantors being the Academy body corporate and myself.

A main driver in spinning out was to have control of our budget and the flexibility to spend where we think best. In our time at RDE and at the PCT we had seen annual budgets become obfuscated by month 2 and meaningless by month 4 with money being moved at will to whichever area the governing organisation had difficulties with. As a social enterprise we immediately had control of the spend and became motivated to be as efficient as possible, knowing that anything we saved would be spent on our service in other ways – not pulled into an NHS black hole.

The success of being in charge of our spending exceeded our expectations many times over in terms of what we were able to achieve. There isn't anything that we previously purchased through the NHS supply chain that we have not been able to get at a better price and often better price and quality when negotiating ourselves. In our hearing aid service where there is a very significant product cost, this was huge – several pounds' reduction in price on each unit, with equipment, marketing and conference places thrown in. The net result of those savings is that we have been able to increase spending on the hearing aid pathways itself. The quality of our NHS product and the time we spend with patients (essential if a hearing aid fit is to be successful) is second to none.

This is a good news story in terms of the service, but the understanding, ownership and participation of staff has been much more of a slow burner. On day one it became clear that there was an expectation that all decisions would now be democratic – subject to a vote. This expectation took some managing. For example, we asked staff what their thoughts would be on our name – we had such diverse and vociferous opinion that various camps were formed! The realisation that we all have our place in the organisation and that in terms of rapid decision making they had to trust the leadership was harsh for some.

Seven years on, many of the staff talk much more about the business and being business-like. For some it is as it has always been – they just want to come to work, do a job and go home. We have a staff council that we have formed after the John Lewis store in Exeter kindly let us shadow their equivalent. This has increasing power. One of the staff council is a member of our board. We offer a bonus for no sickness, all including new starters have access to the NHS pension, and we have life insurance.

An event three years ago precipitated changes in our original Articles. Again we learnt business lessons but we rather fell out with (got let down badly by) the Deaf Academy. It is a story I am better relating verbally but it ended with us having a new chair and also changing the Articles so that all staff have an opportunity to be owners and guarantors. The board now reports annually to an AGM (annual general meeting) of all members and there is facility for an extraordinary meeting and staff to remove the board and MD (managing director) if they have lost confidence. This was a little scary to contemplate at first but I think is evidence of us maturing as an organisation and having confidence in what we are doing.

We have bags of potential. The original hearing aid deal that we had was aggressively usurped by a different supplier who offered us more – crucially, better hearing aids. There is no other audiology service that has spun out and we are seen as movers and shakers but also with suspicion in the industry.

Three years ago we scoped and then developed a business plan for our shop in the Harlequin Centre. Patients told us that in some circumstances they liked seeing us at the RDE but more routinely (where they are Exeter patients, we go to 15 locality clinics around East and Mid Devon) they would like to see us in the centre of Exeter. Some also opined that whilst they accepted that our hearing aids are good that they want the absolute best or smallest and are willing to pay for them. Previously we had sent them on to see someone else but now we worked out how we could see them and direct profit to funding the NHS line. It is also an opportunity for us to sell at very reasonable prices equipment that links with the NHS aids that links them to the speaker, a TV, any Bluetooth device including a telephone. They NHS will never be able to afford to provide this. Despite many claims in advertising, hearing aids work well at seven feet or less from the speaker in relatively quiet conditions – outside of that they are much less effective.

We went to see the CCG (clinical commissioning group) and outlined the plan – they said, 'Great idea but we have no money'. Through contacts we developed a relationship with Social and Sustainable Capital – and after rigorous examination of our plans they lent us the money. It will be paid off this year. So now in a small way we are net importers of funding to the NHS.

Like I said – potential. The CCG invested in us to spin out and should now reap the benefit by expansion wider within Devon and elsewhere. Not so easy. They do not understand what they are commissioning and it seems with even the first contemplation of a tender are not prepared to speak to providers including us as their prime provider. Acute trusts lobby like mad to keep audiology contracts as through gaming of tariffs they believe that they make money. So progress is slower than it might be but we are hanging in there and trying to push the model wherever we can. Something will have to happen in order for the NHS (because it now accepts that there is a minimum level of care through NICE [The National Institute for Health and Care Excellence]) to cope with the growth in elderly population and the need for aiding. There is growing evidence of the contribution of hearing loss incidence and level of dementia as well.

We do have students on placement and they do get a different view of how life can be in audiology through their experience with us. Not sure at this early stage in their careers if they have anything to compare this with, and where we are competing with the private sector this is a chronic shortage. I worry that audiology graduates at such an early stage are making career decisions that are very difficult to get out of – a lifetime of trying to sell commercial hearing aids for high salary. Great to have the money but as evidenced by discussions I am currently having with one graduate of a few years ago, less than rewarding, very focused on one aspect of audiology, and extremely difficult to exit without a change in lifestyle.

Jonathan Parsons, February 2018

***

Given what Jonathan Parsons has described above it is also worth reflecting upon the opportunities provided by social enterprises for a more direct approach in responding to meeting needs.

I would argue that every community, however defined, should have a rapid response action research base, which can take referrals to address issues and problems, and devise ways of addressing those. This would accentuate the value of the network, collaborating with a range of stakeholders who share a value-oriented commitment to creatively providing solutions to problems, both for individuals and collectives. The identification of potential and actual stakeholders plus an understanding of the manner of their contribution to deliberation, assessment, analysis and action could come from such a research base. But, the research base itself should be flexible enough to reflect the specific nature of the task under consideration.

This form of conceptualising intervention as an aspect of community infrastructure building has often been referred to as 'communities of practice', and below I offer as a case study of one such response from Exeter. Before that though I should mention one of my own early contributions to this approach.

In the 1980s while working in higher education in Oxford I was as usual developing my research interests and engaged in course design and delivery. As an aspect of my involvement in youth policy and practice (I designed and taught an undergraduate course on this) I took up the opportunity to work with a local community worker colleague to create 'The Adolescent Network', which sought to bring together a wide range of professional practitioners whose roles involved them in work with young people in the city. We also produced a regular bulletin that was circulated to all interested in such policy and practice issues. The membership of the 'network' covered youth and community workers, social workers, colleagues from the police and probation, adolescent health care specialists and so on. Many of these practitioners had never met before, and often with only a vague idea of what each other actually did. At our regular meetings we all contributed to lively discussions about the state of the city, young people's needs, attitudes and behaviour, the nature of family life and so on. We also discussed and assessed 'our' responses to needs and situations that young people usually encountered in their everyday lives. I also invited expert speakers, often policy makers or report writers, etcetera, who would update us on trends and strategies. The end result was always very illuminating.

Despite the successful role of the small-scale local intervention these provisions in themselves will not be a complete solution to meeting local need because of the continuing influence of national economic and political contexts. The influence of these factors may well be in a significant period of transition to other configurations, and the shifts in discourses on need and how to respond to this may well accelerate change elsewhere. As always, we know that the long-term success deriving from the delivery of key provisions in people's lives remains contingent on a broader range of socio-economic factors. For example there is little benefit to families to be moved into improved accommodation if they cannot afford the rent!

The issues that have been raised above do have a significant role to play in the conceptualising, design and delivery of courses for the education of professional practitioners. What contribution do existing practitioners within an increasing variety of placements make to the thinking about the next generation of professionals? Is there a link between such thinking and their own continuous professional development, and if so how is the process working? If aspiring professionals begin their working life in an organisation delivering social welfare provisions, what experience and ideas do they then take into future formal educational programmes? What is the inter-relation between what they have experienced, reflected upon and understood about role performance and the ideas on offer in a course that purports to raise their educational standards?

As I have argued above, any student in a course leading to professional registration and employment comes with a life history, experience and expertise in some form or other. They are not a tabula rasa even if some course providers might prefer that.

A long-standing colleague who worked in higher education and for the Commission for Health Improvement, and directly for the NHS, commented that

> [My] experience of teaching vocational groups that had or were aspiring to professional status was that although the theoretical areas of knowledge were often seen as interesting in terms of the context the students would then work in (if they continued in Nursing/Teaching/Social Work etc.) their profession, it's relevance to the actual work itself was not seen as 'real'. Greater emphasis was put on the reality of the work placement where 'real' work and experience was gained. It is worth noting that as time goes by more of those who oversee the placement will themselves have been through a degree programme and may be more sympathetic to the ethos and learning that is gained there.
>
> (James 2017)

Part of the problem here in general is our understanding of the often subtle differences between *knowledge* – bodies of theory, theorising, conceptualising, analysis, reflection, recorded accumulated experience, more reflection and so on – and *knowingness*, that low-brow world of the (social) media-saturated and know-it-all solipsistic self, that unmediated state of consciousness created through a lack of exposure to the knowledge self. Part of 'making sense of ourselves' is to address the inevitable encounters with these intellectual banana skins.

My discussion with health and social welfare course providers at a south-west University in January 2018 was interesting in these respects. They certainly believe that the courses they currently offer prepares students for a far more realistic entry point into professional practice than they experienced themselves many years ago. They are rightly concerned that the 'me first' attitude prevalent in society today can create in students a barrier to awareness of social contexts, insight about themselves and others, and developing empathy. However, the courses they offer *do* address these issues, which is a necessary first step towards reflexivity and fitness to practice. An increasing number of their students have educational course placements in the voluntary and charity sectors, and are apparently eager to emulate former graduates in having aspirations to follow them into practice in those sectors. Indeed one colleague at this institution suggested that students and graduates choose to work in the non-statutory sector because they see this as the best opportunity to focus their practice on preventative issues.

As I have suggested above the declining role of the statutory sector as a provider does mean that there is little option for course managers, and students, in where they do placements. It would therefore seem that as more students, in a wide range of health and social welfare roles, experience working in the voluntary/charitable or even social enterprise sectors, the likelihood of them taking up employment there on graduation is much more possible. The relatively smaller scale of non-statutory sector organisations, the closer relationship perhaps to the local 'community' might enable these practitioners to be more autonomous, more themselves. It would seem that an increasing numbers of students might place that meaning on their progress towards professional status.

Of course there are many other factors affecting graduating students' choices, for example pay, working conditions and pensions. There may also be more or less opportunities for employment of a diverse kind because of where a person lives, or wishes to live, or their scope for internal migration to take up a post.

While these shifts in the focus and goals of prospective professionals are happening, the social enterprise movement, to take one example of what I have discussed above as a local basis for community provision, is developing. As already highlighted, the form that any social enterprise takes is extremely variable and dependent on local circumstances and the aspirations of the creators. Initial and later developments are also a key factor, for example with regard to the size and provision range of any one organisation. However, even given all these relevant variables there is a whiff of syndicalism in this relatively new type of provider in our communities. A good deal of what people say about the reasons for setting up an SE, the mission statements, or those who wish to be involved with them, along with many prospective practitioners who choose a professional role in a non-statutory environment remind me of syndicalist ideals: an 'island of socialism in a sea of capitalism'. I look forward to seeing how this works out in the near future, and the extent to which professional practitioners embrace a reforming, even radical social movement ethos. Municipal socialism writ small is a distinct possibility.

Chapter 6

A conclusion

Let me conclude where I started, with John O'Neill and his proselytising role as an advocate for the values of sociology. I have lived *my* sociological life doing the same, and firmly believe that virtue can come from this professional practice linked to reciprocity, meeting needs, compassion, empathy and a humanity coupled absolutely to an endless curiosity for the human condition. This has taken me like all social scientists on a journey of discovery often leading to an understanding of what it is to be human, and attempting to devise ways to explain that to myself and others. My profession may well be as a sociologist, but my vocation is as an educationalist. I would argue that the inter-relationship between the two is what really matters. The collective, collaborative and cooperative quest for 'the good society' is not only linked to virtue through service giving, but also allows the individual to achieve a degree of self-actualisation along the way.

I have also set out in this essay to promote the value of theory, and aimed to *theorise* my way through this essay – cogitating, discussing and writing, thinking-in-action. I have raised the dangers of solipsism amongst social scientists and trust that this has been avoided in this essay. By nature a contrarian, I have attempted to look again in a critical way at the issue of the role of social science in the education of professional practitioners. This has included asking *the* questions: Firstly, do we actually need specifically educated professional practitioners any longer in the realm of health and social welfare in the widest sense of that term? And, do the health and social welfare educationalists actually need us, the social scientists? If they do, what role are we to play in the contextual discussions to curriculum development, and to actual design and delivery?

Given all the changes that have taken place since 1945–48, and the development of the welfare state out of the warfare state, can a case be made to abandon the idea of *professional* practice for both groups of role holders?

Most people reading this book will be only too aware of the social, political and moral pressures that professionals now face. Maintaining integrity in practice has been sorely tested, *is* being tested, and for all manner of reasons we do need a debate about the value of all these interwoven practices, and to devise a workable inter-relation between the various agencies and providers. Policy makers, now and in the immediate future, also need to focus their attention on both the state of the matter now, in the short and longer terms. Not the least of issues here concerns workforce recruitment levels, especially so given the lamentable record of successive governments in the inconsistency of adequately maintaining the necessary staffing levels across health and social welfare. Whatever proposals for health and social welfare provision policy makers decide to put before the UK population and those active citizens of civil society, they *must* consider the role of professional practice education. Policy makers must also engage in the widest possible inclusion of all stakeholders, including trades unions, professional associations in their various forms and the appropriate standards and regulatory bodies. An unequivocal return to peer review is essential for the last of these. Even allowing for the tendency to myopia and political short-termism, policy makers cannot ignore these vital issues.

Certainly what is essential is an unambiguous re-statement of the values that drive us forward and an evaluation of what professional practice in general and social scientific practice in particular can aim to achieve. What can be, and should be, our contribution towards developing ideas and action around what needs to be done to mend a very fragmented and angry society – in fact a society slowly waking up to the extent of centralised and opaque controls over our everyday lives? This is not just political rhetoric on my part, which may or may not be shared by others, but an honest acknowledgement of just how much needs to be done to make redress for so much that has been lost. There is a great deal of very good work being done at the moment amongst most practitioners in diverse settings, but more collective, collaborative and cooperative work needs to be done. And this has to be a democratic process, from each according to their means, to each according to their needs. Do unto others as you would wish for yourself. Professional practitioners in all roles are in an ideal position to expose, discuss and challenge the conventional wisdoms of social relations established over several decades and dominated by the politics of selfishness. This is not an abstract or vaguely idealistic act of exhorting the necessity to address change, this is a very practical argument about what can we *all* contribute, and what are the best ways, including co-working, to do this.

There are of course many barriers to change, not the least being the current 'big business first' values. H. E. Elaine Martin amongst many others researching the development of higher education has commented that working with these paradoxes is demanding:

> [I]n order to work effectively, a series of tensions need to be balanced: vision and reality; individuality and collaboration; reward and accountability; valuing the past and being open to the future.
>
> (Martin 1999 p. 147)

Most certainly a somewhat muted echo of the 1987 tirade by Allan Bloom on the condition of higher education in the USA, *The Closing of the American*

Mind: How higher education has failed democracy and impoverished the souls of today's students. Certainly shades of Richard Sennett here!

Whichever side of that debate we now stand on, I have said a good deal in this essay about the essential and valuable contribution that social science in general has made to a small slice of the human condition, and in particular the preparation of people to become professional practitioners in the field of health and social welfare. Like many other writers of stories reflecting upon the human condition, social scientists have attempted to express their insights into social phenomena and make some sense of it all. Having done this for themselves, social scientists are then able to engage with a diverse range of associates who share those values and debate the best routes to a greater understanding of the issues here, *and* disseminate this to a much wider audience in order to make these processes inclusive and empowering. We are all hostages to fortune in our attempts to do this, and the dustbin of history is full of optimistic but failed attempts to do so even if the value orientation of the social scientist in question coupled with the motivation to go beyond oneself is present. However, the desire to understand and explain is very powerful.

A key issue in this essay focuses on the claim that professional practitioners are well placed to make reasoned interventions into the lives of people, individually and collectively, who have certain needs that should be addressed. The interventions that are, or could, be made are diverse to say the least, and may just extend to giving advice based on *that* practitioner's past experience of similar situations. Drawing on an extensive body of practice knowledge, individual and collective once again, is an obvious example of where interventions can be considered, discussed and assessed, and safeguarding and governance tested. And what about the law? And what about the contemporary discourse about this and that? What current sensation is the news-hungry media promoting *today* for 'our' consumption? And what about the ever-running psychodrama of party politics and government? These are all key contexts to what we do as citizens, as social scientists and as professional practitioners, and we need to talk to each other to make some sense of it all.

In my introduction there is an argument put forward supporting the interrelation between the scope and skill of social scientists, and their role in meeting the needs of people with a very diverse and complex range of life situations. I have argued that social scientists have devised ways to collect and analyse evidence, to theorise about why any social phenomenon may appear to be as it is. What is on the surface? What is hidden away? What are the residual, emergent and latent fixes for our general bewilderment and befuddlement?

All the issues raised in this conclusion underline one of my key concerns in this essay, namely that whatever is decided by whomever, courses for professional practice will have to be designed and delivered. Students will need to be offered an education that broadens their horizons, sets the scene, gives them access to an extensive body of knowledge which they can supplement from their own experience, and prepare for and be mentored towards fitness to practice. I have discussed at length what might constitute that fitness.

It is my hope that social scientists can continue to play a valuable role in the curriculum development processes that will be needed for the education of professional practitioners.

Bibliography

Abbott, J. (1994) *Learning Makes Sense; Re-Creating Education for a Changing Future* Fort Lauderdale: Education 2000

Astley, John (2006) *Professionalism and Practice: Culture, Values and Service* London: The Company of Writers

Astley, John (2008) *Herbivores & Carnivores: The Struggle for Democratic Values in Post-War Britain* London: Information Architects

Berstein, Basil (1970) Education Cannot Compensate for Society *New Society* 26.2.1970

Bourdieu, P. and Passeron, J. (1977) *Reproduction in Education, Society and Culture* London: Sage

Bronner, Stephen Eric (2011) *Critical Theory: A Very Short Introduction* New York: Oxford University Press

Burns, Mel Piece written specifically for this book March 2018

Byung-Chul Han (2017) *Psychopolitics: Neoliberalism and New Technologies of Power* Verso: London

Canna, Crescy (1972) Social Workers: Training and Professionalism. In Pateman, Trevor (Ed.) *Counter Course: A Handbook of Course Criticism* Harmondsworth: Penguin Special

Carr, W. and Kemmis, S. (1986) *Becoming Critical: Education, Knowledge and Action Research* Lewes: The Falmer Press

Coates, Ken and Silburn, Richard (1970) *Poverty: The Forgotten English Men* Penguin Special Harmondsworth

Coats, M. and Tait, J. (2004) *Cobe Web* Open University in House Publication

Cohen, Anthony P. (1985) *The Symbolic Construction of Community* London: Routledge

Community Development Programme (CDP) (1977) *The Costs of Industrial Changer* London: Inter-Project Editorial Team

Craib, Ian (1994) *The Importance of Disappointment* London: Routledge

Damasio, Antonio (2012) *Self Comes to Mind* New York: Pantheon

Deacon, Bob (1983) *Social Policy and Socialism: The Struggle for Socialist Relations in Welfare* London: Pluto Press

Dean, Malcolm (2012) *Democracy Under Attack: How the Media Distort Policy and Politics* Bristol: The Polity Press

Dorre, Klaus et al (2015) *Sociology, Capitalism, Critique* London: Verso

Duncan, Graeme (1993) *Democratic Theory and Practice* Cambridge: Cambridge University Press

Elliott, Anthony (2001) *Concepts of the Self* Cambridge: Cambridge Polity Press

Erickson, Frederick (2004) *Talk and Social Theory* Cambridge: Cambridge Polity Press

Everitt, Angela and Hardiker, Pauline (1996) *Evaluating for Good Practice* London: Palgrave Macmillan

Everitt, Angela et al (1992) *Applied Research for Better Practice* London: Palgrave Macmillan

Feuerstein, R. R., Hoffman M., and Miller, R. (1980) *Instrumental Enrichment* Jerusalem: The Feuerstein Institute

Field, John (2003) *Social Capital* London: Routledge

Garfinkel, Harold (1967) *Studies in Ethnomethodology* Englewood Cliffs: Prentice-Hall

Gerth, H. and Mills, C. Wright (1954) *Character and Social Structure: The Psychology of Social Institutions* London: Routledge & Kegan Paul

Gibson, James (1979) *The Ecological Approach to Visual Perception* Houghton Mifflin Boston, U.

Giddens, A. (1991) *Modernity and Self-Identity* Stanford: Stanford University Press

Goffman, Erving (1959) *The Presentation of Self in Everyday Life* Harmondsworth: Penguin

Goffman, Erving (1968) *Asylums* Penguin Books Harmondsworth

Goffman, Erving (1968) *Stigma* Penguin Books Harmondsworth

Goldthorpe, John H. et al (1969) *The Affluent Worker in the Class Structure* Cambridge: Cambridge University Press

Gomm, R. and Davies, C. (Eds.) (2000) *Using Evidence in Health and Social Care* London: Sage

Gopnik, Alison (2007) How Ruthless Should We Be About Advancing the Welfare of Our Children *The Guardian* 27.8.2007

Gray, Collette and MacBlain, Sean (2012) *Learning Theories in Childhood* London: Sage

Hall, Stuart, Held, David, and McGrew, Tony (1992) *Modernity and Its Futures* Cambridge: Polity Press

Hannan, Andrew and Silver, Harold (2000) *Innovating in Higher Education: Teaching, Learning and Institutional Cultures*. The Society for Research into Higher Education and The Open University Press Buckingham

Halsey, A. H. (2004) *A History of Sociology in Britain* Oxford: Oxford University Press

Halsey, A. H. et al (1980) *Origins and Destinations* Oxford: Oxford Clarendon Press

Hamilton, Paul (1996) *Historicism* London: Routledge

Harloe, Michael et al (Eds) (1990) *Place, Politics and Policy* London: Unwin Hyman

Hart, E. and Bond, M. (2000) In R. Gomm and C. Davies (Eds) *Using Evidence in Health and Social Care* London: Sage

Hills, Rachel (2018) *Youth and Community Work Course, Ruskin College, Oxford* Case study piece written for this book

Hind, Dan (2010) *The Return of the Public* London: Verso

Hoggart, Richard (1957) *The Uses of Literacy: Aspects of Working-Class Life: With Special Reference to Publications and Entertainments* Harmondsworth: Penguin

Honey, P. and Mumford, A. (1986) *The Manual of Learning Styles* Bradford: MCB UP Ltd

Illich, Ivan D. (1971) *Deschooling Society* Harmondsworth: Penguin

Integrating social prescribing and community building. Integrated Care Exeter. www.wellbeingexeter.co.uk

James, David (2017 December) Professional Education Personal memo to me

Jaques, David, Gibbs, Graham, and Rust, Chris (1990) *Designing and Evaluating Courses* Oxford: Polytechnic

Jarvis, Peter (1983) *Professional Education* Beckenham: Croom Helm

Karabel, Jerome and Halsey, A. H. (Eds.) (1977) *Power and Ideology in Education* New York: Oxford University Press

Kings Fund Website (2017) *What Is Social Prescribing*

Knowles, Malcolm (1973) *The Adult Learner: A Forgotten Species* Houston: Gulf Publishing Co

Kynaston, David (2008) *A World to Build* London: Bloomsbury
Lacey, Barry (2018) *Journeys Without End* Case study piece written for this book
Lacey, Barry (2018) *Journeys Without End* Piece written for this book
Lindblom, Charles E. and Cohen, David K. (1979) *Usable Knowledge: Social Science and Social Problem Solving* New Haven: Yale University Press
London Edinburgh Weekend Return Group (1980) *In and Against the State* Harmondsworth: Penguin
Mann, Chris Piece (2018 March) written specifically for this book.
Marcuse, Herbert (1964) *One Dimensional Man* Boston: Beacon Press
Martin, Elaine (1999) *Changing Academic Work: Developing the Learning University* Buckingham: The Open University Press
Merriam, Sharan B. (Ed) (2001) *The New Update on Adult Learning Theory* San Francisco: Jossey-Bass
Mills, C. Wright (1959) *The Sociological Imagination* New York: Oxford University Press
Needham, G in Gomm and Davies 2000
Oliver, Billie (2005) *What's in a Name? Professional Identity in Transition* Conference paper July
O'Neill, John (1972) *Sociology as a Skin Trade* London: Heinemann Educational
Pahl, Ray (2005) *Are All Communities in the Mind?* The Sociological Review Volume 53 issue 4 November 2005 Wiley U.K.
Packard, Vance (1960) *The Waste Makers* Harmondsworth: Penguin
Pallitt, Christopher et al (Eds) (1979) *Public Policy in Theory and Practice* Sevenoaks: Hodder & Stoughton and The Open University Press
Parsons, Jonathan (2018) *Chime* Case study piece written for this book
Riesman, David (1964) *Abundance for What? And Other Essays* London: Chatto & Windus
Ryan, Alan (1970) *The Philosophy of the Social Sciences* London: Palgrave Macmillan
Sandford, Jeremy (screenplay) and Loach, Ken (director) (1966) *Cathy Come Home* BBC1 Wednesday Play
Samuel, Raphael (1994) *Theatres of Memory. Volume 1 Past and Present in Contemporary Culture*, Verso London
Schon, Donald A. (1987) *Educating the Reflective Practitioner, Toward a New Design for Teaching and Learning* London: Jossey-Bass
Sennett, Richard (1998) *The Corrosion of Character* London: W. W. Norton
Sked, Alan and Cook, Chris (1993) *Post-War Britain: A Political History 1945–1992* London: Penguin
Smith, Mark and Jeffs, Tony (1996) *Informal Education: Conversation, Democracy and Learning* Education Now Publishing Cooperative Ltd. Nottingham
Smith, George et al (2014) *Social Enquiry, Social Reform and Social Action* Oxford: University of Oxford, Department of Social Policy and Intervention
Steiner, C. and Perry, P. (1997) *Achieving Emotional Literacy* London: Bloomsbury
Stevenson, Olive (1976) The Development of Social Work Education. In A. H. Halsey (Ed.) *Traditions of Social Policy* Oxford: Basil Blackwell
Tawney, Richard (1961) *The Acquisitive Society* Collins (Fontana Library) London
Thompson, S. and Thompson, N. (2008) *The Critically Reflective Practitioner* Basingstoke: Palgrave Macmillan
Tod, Tim (2018) *Youth Work in School* Case study piece written for this book
Townsend, Peter (1979) *Poverty in the United Kingdom* Penguin Harmondsworth
Urry, John (1990) *Conclusion: places and policies* Chapter 9 in Harloe 1990

Wilkinson, R. and Pickett, K. (2009) *The Spirit Level: Why More Equal Societies Almost Always Do Better* London: Allen Lane

Williams, Raymond (1965) *The Long Revolution* Harmondsworth: Penguin

Young, Michael F.D. (Editor) (1971) *Knowledge and Control* Collier MacMillan West Drayton

Zuber-Skerritt, Ortrun (1992) *Action Research in Higher Education: Examples and Reflections* Falmer: Kegan Paul

Index